D1285213

FARM TOOLS

through the Ages

Also by Michael Partridge
Early Agricultural Machinery

FARM TOOLS
through the Ages

Written and illustrated by

Michael Partridge

PROMONTORY PRESS

631.3
P258f

Copyright © 1973 by Michael Partridge
All rights reserved
Library of Congress Catalog Card Number: 75-12296
ISBN 0-88394-035-3
Designed by Behram Kapadia
Printed in the United States of America
Published by arrangement with New York Graphic
Books, Boston

Preface

The evolution of modern farm machinery, from the simple tools used by primitive people, was not one of smooth progress. It often took place in isolation or against a background of scepticism from farmers and labourers alike. Progress was mostly achieved on a basis of trial and error by enthusiasts whose efforts, perhaps over half a lifetime, were lost unless they could cajole neighbours into adopting the improvements.

In very early times we can be sure that the digging stick was used literally to scratch an existence from the soil. Later a huge wooden spade, called the caschrom, gained a far deeper penetration and caused some inversion of the soil. The first crude ploughs were pushed by men and later by domestic animals. They simply made shallow ruts across the soil in preparation for the seed that was to be hand-broadcast. The Romans had a variety of ploughs, a reaping machine and various implements for cultivating soil, but what little farming knowledge was absorbed by the Britons under Roman rule was largely forgotten when the legions withdrew. Saxons and Normans cleared large tracks of land by fire and the heavy single furrow plough, drawn by a powerful team of oxen, was a particular advantage. A regular system of crop rotation became established and the farmer's scope was later increased with the introduction of new vegetables. Responsibility for plough manufacture lay entirely with blacksmith and carpenter when Walter Blith laid down requirements of good plough design and thus provided a foundation for others to build upon. The eighteenth century saw a remarkable advance in the form of Jethro Tull's practical seed drill and his book, *Horse Hoeing Husbandry*.

Scientific investigation was soon to be stimulated by the growing number of agricultural societies in Britain, and, to no lesser degree by the travels and writings of the celebrated agriculturalist Arthur Young.

The ancient flail was gradually replaced by threshing and winnowing machines and by the mid-nineteenth century various reaping machines were available. Ideas and improvements had been spread abroad from Europe to America and Australia but the new settlers were forced to develop implements capable of tackling the most difficult terrain. American and Australian inventors were particularly responsible for improved harvesting machinery during the nineteenth century, whilst at home in England there was a great deal of experimenting with steam for the purpose of cultivation, using huge ploughs and diggers. The horizon of agricultural endeavour was greatly extended in Victorian times by the audacity of farmer-engineers but few of their gargantuan devices were ever proven practical. John Fowler surpassed the efforts of most with his steam ploughing tackle. His sets were taken all over the world and must have seemed the ultimate method of land cultivation at a time when the first oil-fired tractors appeared on the farming scene. Much valuable information regarding the invention and use of farm tools and implements can be gleaned from early writings, but the descriptions become more precise in the educational treatise that began to flow forth during the eighteenth century. They provided the farmer with much of his instruction and also recorded the efforts of the many ingenious gentlemen working towards agricultural improvement. It is pages such as these that provide much of the knowledge contained herein, and if I have followed their lines too closely without acknowledgement it is unintentional. Also the museums specialising in agricultural history have afforded me a great deal of familiarity with the simpler tools so often neglected by writers. I have endeavoured to render both verbally and pictorially an accurate description of the evolution, use and construction of the tools and machines employed by land owners over the centuries to modern times.

Ancient and modern almost always overlap in rural areas, and nowhere is this more apparent than in corners of the British Isles, where people tend to retain customs and traditional practices with fierce independence. If one wanders far enough it is possible to see these rural craftsmen working at coracle-making, weaving at hand looms, quilting, cheese-making, cider-making, peat-cutting and more. They are happy people full of the art and craft that lingers after centuries to remind us of the time when life was less complicated. In these out of the way places it was possible until quite recently to see a crofter using the caschrom and a farmer broadcasting his seed by hand, whilst nearby, in complete contrast, the same work was completed much faster and more thoroughly with the latest machinery. Farming nowadays is highly intensified. Soil and crop processing has become so mechanised that much of the work has to be done on a contract basis and there seems no place for old-fashioned ways. Modern farmers handle their automatic implements with the fingers of a skilled mechanic, but still retain that inherent knack of reading weather signs, handling animals and generally conducting business in a quiet, practical manner as did their forerunners. Nowadays there is a growing nostalgia for the past and a desire to preserve as much as possible for future generations. Already an enormous amount has been achieved. A great deal of information and many examples of equipment representing different activities from the past have been successfully collected together, and it is no wonder that many of them are the tools connected with production and preparation of foodstuffs. All the better if these 'seemingly mysterious devices' could be seen in use. Perhaps this is a plea to preserve some full-size working farms where in due seasons seed would be sown by hand, corn harvested by sickle and then threshed by flail; where horses would be kept in number to haul the implements; and the blacksmith's, wheelwright's and other country crafts kept very much alive.

ACKNOWLEDGEMENTS

The author wishes to acknowledge the valuable assistance of:

Miss Anthea V. Driver, Rutland County Museum; Dr G. E. Fussell; Mr Andrew Jewel, The Museum of English Rural Life, Reading; Miss N. W. Rhodes, The University of Nottingham; The National Institute of Agricultural Engineering; The McCormick Historical Society, Chicago; Miss Veronica Sharman; Mrs Kay Savage.

Wherever possible the drawings were completed from the collections of the Rutland County Museum and the Museum of English Rural Life, Reading.

Contents

Illustrations

'The introduction of new implements into a district is often a matter of the greatest difficulty, owing to the ignorance, the prejudices and the obstinacy of farm servants and labourers. Many farmers therefore, very absurdly retain their old implements, though convinced of their inferiority, rather than sour the temper of their labourers by attempting to introduce new ones. In many cases they have succeeded, by attention, by perseverance and by rewarding their farm servants who have been induced to give the machines a fair trial.'

FROM *Code of Agriculture* BY SIR JOHN SINCLAIR, *1817*

I

Land drainage

European farmers in particular have the problem of wet land to contend with, and since the earliest times have been required to expend a considerable amount of time, labour and ingenuity on drainage, in order to render their land suitable for cultivation. The Romans were first responsible for improving some parts of Britain and by laying down open drains and ditches to collect and carry away excess water, they converted wet marshy land into fields suitable for growing crops. They may also have introduced covered drains, which were known in classical times, and consisted of stones or brushwood laid along the bottom of a deep trench covered with soil. This served to direct a flow of water to an outlet or main receiving drain, but such constructions were short-lived as they quickly became blocked with fallen soil or collapsed. These methods were continued by the native Celts after the Romans left Britain, and by the Middle Ages ditches and open furrows were still the main forms of land drainage.

The Saxons used deep furrows as a means of directing surface water, and also to define the boundary between each man's strip or acre of land. Using a heavy wooden plough drawn by oxen, they formed a ridge straight down the middle of a strip by leaning two furrow slices up against each other, and then proceeded to complete the work by ploughing up and down on both sides of the ridge leaving all of the furrow slices leaning in towards it. The same process was repeated on the adjacent strip and when the two were completed there were two ridges, with a broad furrow lying midway between them. But continued ploughing, year after year in exactly the same manner with no variation in the position of the strips, caused the ridges to become gradually higher and the furrows between them deeper, so that in some instances they were deep enough fully to contain a man. The furrows were vitally important and served for drainage. They received rain water from the ridges on either side, but in the process a good deal of valuable top soil and nourishment must have been washed away.

The open field strip system gradually gave way to enclosed farming after about 1200. The change over was slow but enclosed farming gradually increased in one district and another throughout England over the following centuries.

Then the farmer was left to his own devices and his success was largely dependent upon his own ingenuity, knowledge and skill. The more enterprising ones set about drainage in a thorough manner and laid systems of open water courses and ditches. By the sixteenth century some farmers were excavating deep pits or 'soughs' about their land, which they filled with brushwood and covered with turf to soak up water from the surrounding area. At that time open water furrows were used rather than covered drains, and agricultural writers began to advise on their arrangement and construction. Furrows were also used to direct water in measured amounts from brooks and rivers, across the surface of dry meadows. This **meadow watering** was executed early in the winter season and the water was allowed to remain on the meadows for two or three months, after which other furrows were made to carry it away.

Walter Blith was probably the first writer to record the use of instruments especially intended for drainage work, and in his *English Improver Improved*, 1653, he described the horned draining spade, various gouges, and a trenching plough, all of which were used to manufacture water furrows. The **trenching plough** contained a single coulter below its beam and simply cut each side of the intended trench; the soil between the two cuts was then removed by hand digging. In the early years of the following century, the Cambridgeshire draining plough emerged in the Caxton area. It replaced Blith's plough and also formed the basis of most later designs. It was a common pattern of plough but was furnished with two parallel coulters and a broad flat share and was capable of excavating a trench 18 in. wide and 12 in. deep in a single operation. The Royal Society of Arts encouraged the development of an improved draining plough, and in 1760 they awarded prizes to two inventors, but neither of them completely overcame the difficulties encountered with this kind of work and their implements required exceptionally large teams of horses on wet heavy land.

Whilst the old brushwood drains were continued, some new types of underground drain began to find favour. Circular channels similar to the later mole drains were formed beneath the surface by a method known as rowl or **plug drainage.** Also in many districts huge quantities of small stones were used partly to fill trenches or, alternatively, hollow channels were formed below ground by the use of large flat stones. Such forms of land drains remained popular until the nineteenth century when specially designed earthenware drain tiles became available.

By the beginning of the nineteenth century, Joseph Elkington had established himself in the forefront of land drainage in the midland counties. His idea of intercepting water at its point of origin quickly caught on and he was sponsored by the Board of Agriculture further to perfect his work. An advertisement to the first edition of *The Mode of Draining Land*, an 1801 publication which described his manner of work, read:

The Board of Agriculture had hardly been established before it received intelligence from various parts of England of the singular success with which Mr Joseph Elkington, a Warwickshire farmer, practised the art of draining land; the publication or discovery of which was represented to be one of the greatest means of promoting the improvement of this country, that could be suggested. It may be of sufficient interest to mention that in consequence of a motion made by its President on the 10 June 1795, the House of Commons voted on address, that his Majesty would be graciously pleased to give directions for issuing to Mr Joseph Elkington, as an inducement to discover his mode of draining, such sum as his Majesty in his wisdom shall think proper, not exceeding the sum of £1000 sterling.

Using his patent boring irons, Elkington penetrated vertically into the earth, or horizontally into hillsides and released pent up water or redirected springs into stone-filled trenches.

The **mole plough** first came into use towards the end of the eighteenth century, and its hollow subterranean bore effectively relieved the problem of clayey land that normally retained the winter rainfall until the very late spring. Its design has changed only slightly to the present day, but the working principle has never been altered or improved upon. It did, however, capture the imagination of John Fowler who adapted it to steam haulage some years after his successful attempt to pull a string of wooden pipes into the earth behind the mole.

Other inventors had also been at work on various devices for draining farm land. An enormous amount of interest was aroused by a Scotsman, James Smith of Deanston, who perfected his **subsoil plough** during the 1820s and using it to break up hard packed subsoil along with a simple arrangement of thorough drainage, improved his poor farm land beyond recognition. His neighbours were greatly impressed with the result, and the news soon spread throughout Scotland and England. Other farmers began to imitate his methods, commonly called 'Deanstoning', and they quickly saw an improvement in their own land.

The manufacture of cylindrical baked clay draining pipes was achieved by John Reade early in the nineteenth century and Thomas Scragg was first to manufacture them in quantity by machine in the 1840s. Attention was then turned towards lightening the labour of trench cutting. Rotary diggers, powered by capstan and horses, were tried out but the majority of them did not develop beyond the experimental stage. Steam-driven machines that opened a trench, placed pipes in position, and then replaced the soil, were featured at Agricultural Shows towards the turn of the century. But such things were not for the ordinary farmer, and if any were ever purchased, which is doubtful, owing to the considerable size and weight, it was by drainage contractors.

Water meadows are still to be found alongside some rivers in chalky districts of England and on the Continent, but the majority have fallen into disuse owing to the high cost of maintenance and repair. The practice of meadow watering was widely used during the eighteenth and early nineteenth centuries, and was designed to place a large area of land under water where it was protected from the harsh effects of frost during the winter months. During November, water was directed off from the main flow of a river and released through hatches into a series of interlacing drains. It was carried over the full extent of the waterside meadows by shallow 'leader' drains and returned to the river lower down by means of deeper drains. The meadows remained flooded for the whole of December, in January they were allowed to dry out twice, then in February the water was taken off by day and flooded by night. It was finally drained away in March and resulted in an early bite of grass for the benefit of ewes and lambs, and later in the year a superior quality of hay. English counties that excelled in this practice were Wiltshire, Hampshire, Gloucestershire, Worcestershire, Devonshire and Berkshire. Dorset once possessed 6,000 acres of water meadow, whilst there were many other riverside meadows that flooded of their own accord. In England, water meadows were generally leased by smaller farmers from Lady Day to May Day.

Types of Land Drain

Open drains called dykes or ditches, were important in the early stages of land drainage. The Romans introduced them into Britain and they were later subscribed to by Fitzherbert, in his *Boke of Husbandrye, 1523*. He recommended them above stone and faggot drains and they remained the chief method of land draining until the end of the seventeenth century. But owing to the constant need for repair and scouring they were generally replaced wherever possible by underground drains. However they continued to be most practical for drying the surface of mountain pasture land and in high country where the cost of laying underground drains would not be returned.

Brushwood drains consisted of a narrow V-shaped trench, filled to a depth of several inches with alders, willows, limes, rushes or other suitable material. The faggots were slightly overlapped with the butt ends placed towards the outfall of the drain. Sometimes a layer of straw, twigs, or stones was laid on top of the brushwood before the soil was replaced and an effective drain thus completed. Larch poles were very often laid along the bottom of a trench to form a crude sort of conduit, but it easily became blocked by vermin or fallen earth. Bush and faggot drains were used from Roman times and could be found especially in those areas where stones were scarce. They were generally replaced by pipe drains in the nineteenth century.

Stone drains. In the seventeenth century Walter Blith recommended drains 3 to 4 ft. deep in which the lowest layer of faggots was covered by turf, then by 15 in. of stones. The remainder of the trench was filled with soil. Stone drains were subsequently made wherever a sufficient quantity of small stones, broken stones or rubble was available. Normally the lowest 12 in. of a 3 ft. deep trench was filled with such material and it admitted a direct filtration of water from the land above. The stones composing a drain were ideally less than $2\frac{1}{2}$ in. in diameter which was ascertained by the stone breaker who passed each piece through an iron ring of that diameter.

Box drains were constructed with stones so arranged in the bottom of a trench as to form a hollow channel about 6 in. square. The flattest stones were laid along the bottom and small ones stood upon edge, along the sides. The channel was covered over with more flat stones and the trench

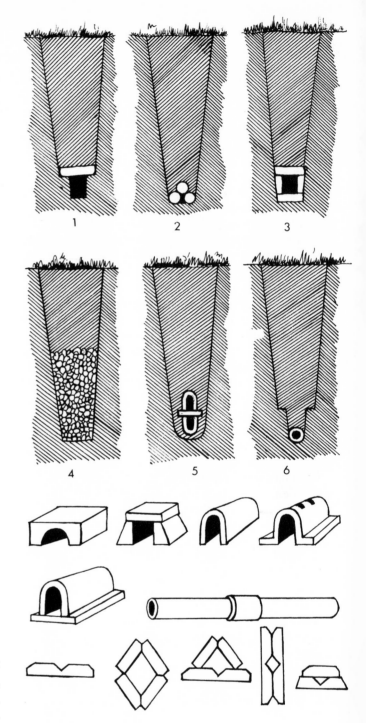

Cross sections of
1 Shoulder or wedge drain
2 Brushwood or pole drain
3 Box drain
4 Stone drain
5 Horseshoe tile drain
6 Pipe drain
Below: A variety of nineteenth-century draining tiles

above it filled with soil. Such drains were probably first made in the first half of the seventeenth century and used until flat stones were replaced by the more manageable eighteenth-century drain tile.

Wedge drains employed in areas where stones were scarce, required shoulders or ledges to be left part way down the trench. This was achieved by the narrowest draining spade which formed a wedge-shaped cavity in the bottom of the trench. If they were available, flat stones were laid across, resting on the ledges at each side, leaving a vacant space below, but more often, turf, with the grassy side undermost, was used instead of stones. This method of drain construction was probably developed during the late eighteenth century and continued as a cheap substitute for hollow tile drainage.

Tile drains. The earliest pattern of land drainage tile was a simple arch or tunnel, called the horseshoe tile. They were made with burnt clay, placed end to end along the bottom of the trench and covered over with soil. The improved pattern of this tile had a separate flat or sole of the same material, and then each horseshoe tile was placed resting upon two adjoining soles to lessen the danger of sinking. Drain tiles were made and used extensively during the latter part of the eighteenth century. A selection of nineteenth-century drain tiles are illustrated.

Pipe drains. Cylindrical drainage pipes were first manufactured by John Reade, a gardener of Horsmenden, Kent, at the beginning of the nineteenth century. He made them by folding a thin sheet of clay, about 12 in. square, around a wooden shaft, but distortion during the firing usually resulted in a slight taper towards each end and a slit along the length of the cylinder. Cylindrical pipes were laid in the ground with and without the use of joining collars, and they superseded all other forms of drainage tile. They are still laid extensively but the trenches are opened and closed by mechanical diggers.

Plug drainage. Before the mole plough became generally available, some farmers constructed temporary drains beneath their land by a method first known as rowl and later as plug drainage. It became popular in the second half of the eighteenth century, and was cheaply executed as no expensive implements, heavy horse teams or costly materials were required, whilst the resulting hollow drains 20 in. or more below the surface of the soil were efficient and durable. In this type of drainage, after the first two spits had been removed the digging

Plug drainage

was completed by an instrument called the **bitting iron** which formed a neat and narrow bottom to the trench. Any crumbs of soil or small stones were next removed from the bottom with a scoop, and a wooden mandrel was then brought into use. It took the form of a flexible pole formed by a number of cylindrical pieces of wood, each about 12 in. in length and about 3 in. in diameter, fastened together end to end by iron links, with a length of chain attached at the end. The mandrel was placed in the bottom of the trench, and the soil previously excavated was thrown back and pressed down on top of it. The mandrel was then drawn along the trench and pulled free of the soil packed above it, by means of an iron bar which was inserted through a link in the chain and acted as a lever. When all but the last two sections of the mandrel had been exposed, that portion of the trench then above it was filled with soil, and the mandrel was drawn along to the next stage. So the work proceeded along the entire length of many parallel trenches and resulted in a network of continuous hollow drains through which water could flow quite freely. Sometimes holes about 1 in. in diameter were made from the surface of the land to the drain channel below. A long punch was hammered down whilst the mandrel was still in position. The ideal situation for plug drainage was through clay beneath pasture land; in other places it was likely to collapse, due to the weight of horse teams or the vibrations of cultivation implements, moving on the surface above. Neither could the **subsoil plough** be admitted to work where Plug draining was installed. By the middle of the nineteenth century farmers had turned to cylindrical clay pipes as a more permanent form of drain and to the mole plough as a faster means of completing temporary hollow drainage.

Hand Tools used in Drainage Work

Only a few hand tools were employed in the formation of drains and it was for the cutting of drainage trenches that special tools were designed during the seventeenth century. Some were retained for the purpose of pipe laying until digging machinery was perfected in the present century.

Draining spades were usually of three different sizes, gradually diminishing in width, and each was suited to its particular section of the work, which resulted in a smooth sided V-shaped drain some 3 ft. in depth, from which only the necessary amount of soil had been excavated. The spade used to cut out the first spit was 12 in. wide, and was followed by a second spade 8 in. wide. A long, narrow spade, commonly called a bottoming tool, was employed for taking out the last narrow spit where the draining pipes were seated. Its blade tapered from 6 in. to 3 in. across its cutting edge, and necessitated a foot tread at the rear of the blade in order for the worker to exert the required pressure for digging. Each blade was forged in iron and was attached by means of a long split socket to a wooden helve of ash or oak. Narrow blade drain spades are still available.

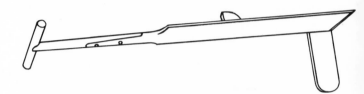

Narrow drain spade
Bitting iron for use with plug drainage, *c.*1870

The bitting or grafting iron was the instrument used from the late eighteenth century for taking out the deepest spit in **plug drainage.** Its iron blade was usually $1\frac{3}{4}$ in. in width across the cutting edge, the sharp triangular projection on the side of the blade 6 in. in length, the distance between the foot tramp and the bottom of the blade 18 in., and the iron handle about the same as the trenching fork. In the formation of plug drains, the top spit was removed with a common spade and laid to one side, then the second spit was taken out with a smaller sized spade and laid on the opposite side of the trench. The sides were sloped inwards to such an angle as left the bottom almost the same width as the blade of the bitting iron, by which a lower portion of earth $1\frac{3}{4}$ in. wide and 15 in. deep was extracted. This tool was laid aside when plug drainage lost favour to mole drainage early in the second half of the nineteenth century.

Trenching spades were probably in use before the seventeenth century, to cut water courses on moor and meadow lands. The worker pushed the spade in front of himself using it as a gouge, with the handle almost flat along the ground. The necessary

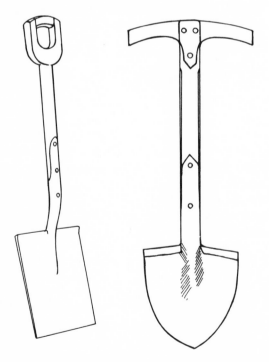

When digging a 3-ft. drain the top spit was taken off with a common spade, then narrower spades were used
Drain spade used in Yorkshire during the nineteenth century

Seventeenth-century farmer releasing water with the horned trenching spade

force was applied by the arms and shoulders with additional pressures being added on the end of the handle by the haunches or thighs. Its iron blade was slightly rounded and furnished on each side with a sharp-edged wing or horn-like projection to cut through turf, roots and other growth. An enormous amount of skill must have been required in order to remove one spit after another until a deep water course had been formed. Such spades were used in England until about the end of the eighteenth century when the labour was lightened by improved draining spades.

The foot pick was especially made to disrupt hard, stony ground that was difficult to pierce with a spade, and for occasions when the level was too deep or obstructed to wield the hand pick with safety. The illustrated pick is a descendent of the ancient mattock and was made during the nineteenth century. It was between 3 ft. 6 in. and 4 ft. in length, having a pointed blade that closely resembled a scimitar. An iron cross handle was provided at the top-most part of the blade, on to which the user exerted the pressure of both hands. A tramp was provided half-way down the blade and at right

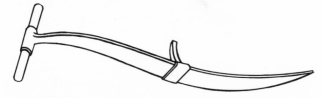

Foot pick, *c.*1870

angles to it, so that pressure could be applied by one foot of the user wherever necessary. When at work, the user first stabbed the point of the pick down into the ground, then gained greater depth by moving the blade with a screwing action whilst pushing down on the tramp with his foot. When sufficient depth was achieved, the user would pull the handle back towards himself and the rising blade would throw up the soil in a loosened state. The pneumatic drill is now used in preference to the foot pick for road building and trenching across difficult terrain.

The pick, or pickaxe, replaced the foot pick for digging hard stony land during the nineteenth century. It is more like a mattock, the blade having a point at one end and a chisel or axe at the other.

Trenching forks were of assistance when the worker wished to loosen or remove subsoil from

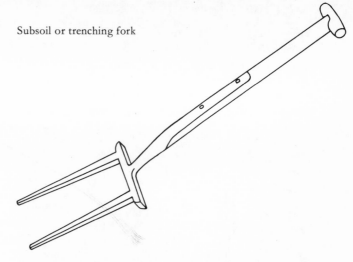

Subsoil or trenching fork

the bottom of a drainage trench. The topsoil had first to be removed with spades, and the subsoil exposed in order for the forks to be used. They were made with either two or three prongs, and the most appropriate number was selected in relation to the type of subsoil encountered. It was found that the three-pronged fork worked well with soft subsoil, and two prongs where flints were numerous. Almost 4 ft. in length, with a wooden helve of oak or ash wood, the trenching fork had two or three prongs, each about 15 in. in length, tapering from $1\frac{1}{2}$ in. square just below the foot tread to a rounded point at the end. Such forks were also used for breaking up tightly packed subsoil surrounding land drains so that surface water could readily pass through it. The Romans used forks for digging but heavy trenching forks were probably developed during the eighteenth century.

The drain stone rake had three long tines, 1 in. square at the top, tapering to a sharp point and set at right angles to the helve. It was used to arrange large and small stones when roads or land drains were being constructed. The Romans had rakes for use in cultivation, but it was probably not until the eighteenth century that they were made for the purpose of moving stones.

The drain scoop was used to smooth off the bottom of a newly made land drain or to remove loose earth left there after digging. It consisted of a flat iron blade about 4 in. in width and 12 to 18 in. in length, sharpened across its front end and raised up for an inch or more along each side. The neck, joining the blade to the helve, was curved upwards, so that when the blade was working horizontally along the bottom of the drain, the 6 ft. long helve came up at a convenient angle to the workman's hip.

He stood astride the drain and pushed the tool along in front of him. Scoops and ladles, made by blacksmiths, were meant to last a lifetime and suited the user like a sickle or scythe. During the eighteenth century they replaced a crude type of wooden scoop carved from a tree branch growing to convenient shape.

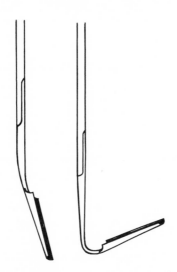

Nineteenth-century drain scoop and ladle

The drain ladle was used before tile or pipe laying commenced in order to remove any mud or water present in the bottom of the drain. A wooden helve was fixed at an angle of 45 degrees to a shallow iron trough which was left open along the far edge to catch up liquid. The helve was made long enough for the labourer who stood astride the drain to reach the bottom easily, and by means of a backward motion scoop out any unwanted slurry and place it to one side. A variety of scoops, each with a different size of trough, were kept at hand by the drain constructor. Drain scoops and ladles are still used to a limited extent in drainage work, and date from the seventeenth century.

The pipe layer designed during the early nineteenth century enabled the workman to position draining pipes end to end without his having to go into the trench or stretch to the bottom with his

Drainpipe layer

arms. Its wooden handle was usually about 6 ft. in length and finished at the bottom end with a 9 in. iron prong, placed at a right angle. The workman inserted the prong of his layer into a new pipe which he then lowered into the bottom of the trench and adjusted it onto the end of those already laid. This instrument is still employed by farmers when laying drainage pipes.

The gripping spade was an effective instrument of obscure origin, used in the north of England for cutting through heather and roots before a ditch was excavated. The helve was short, about 2 ft. in length, provided at the top with a long cross-handle and at the bottom attached into the socket of an iron blade, the shape of which can be seen in the illustration. The blade was sharpened down the length of its longest side and was used in the manner of a saw to cut through top growth along each side of the intended ditch. Once the blade had been inserted, the worker continued his sawing action along that side of the work and then returned cutting the other side parallel to it. The turf between was then removed with a fork, and the soil with a common ditching shovel.

Gripping spade

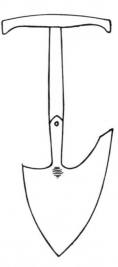

Ditching spades such as those illustrated were commonly used until the latter part of the nineteenth century for cleaning out ditches and ponds etc. They were made wholly of wood, generally ash, because iron would quickly rust away if used continually in water. The helve and blade were shaped from one piece of wood about 4 ft. in length, whilst an easily replaceable iron-shoe afforded the necessary protection to the end. This

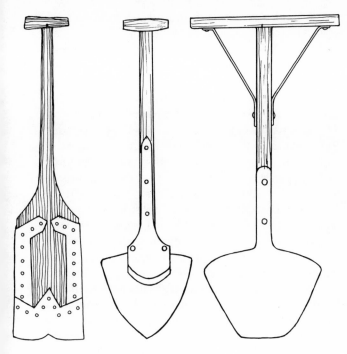

Ditching spades

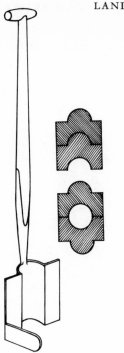

Peat tiles formed by Mr Calderwood's patent spade (*left*)

type of spade probably evolved during the Middle Ages. Also illustrated are ditching spades with iron blades. These were blacksmith-made and more general purpose than the former wooden spades.

The ditching shovel, factory made in the nineteenth century was common to all parts of England. It was furnished with a helve about 2 ft. in length and was convenient for working in confined spaces. Its iron blade was usually 12 in. wide across the top edge where it was folded over slightly to form a suitable foot rest, and 12 in. deep, curved to a point, and sharpened around the edge for hacking through undergrowth, rushes, roots or weeds, as well as for clearing the sides and bottoms of ditches. This pattern of shovel and also the gripping spade, probably came into use during the eighteenth century and have continued, especially in the north of England, until recently.

The ditching shovel with sharpened edges for cutting through thick undergrowth etc.

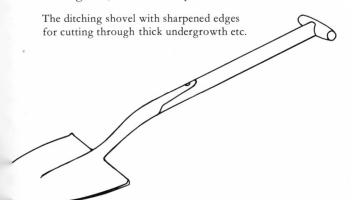

Mr Calderwood's peat tile spade. In areas where peat was plentiful, it was customary during the early nineteenth century to use that material for the construction of land drains. The peat had first to be formed into tiles which were cut by a particular pattern of spade, invented in the early nineteenth century by a Scotsman, Mr Hugh Calderwood. The cutting blade of his spade was made to the shape of a half cylinder, with flat projections on either side, one of which retained a flange 9 in. in length at right angles to the face of the blade. The tiles were formed by the insertion of such a blade into blocks of peat, and as shown in the illustration, the top and bottom surfaces of the finished tiles were identical to the shape of the spade. After being dried out in an oven, or by the sun, the tiles were placed end to end along the bottom of a drainage trench with the same pattern of tile upturned and placed on top to form a continuous circular passage, which was afterwards covered over with soil. These tiles were quite durable, even in boggy land, and in some areas where peat was bountiful they remained popular until the twentieth century. The peat tile spade was necessary until then.

The stone screen was useful for separating a cartload of stones into various sizes and would be constantly required when the farmer was laying his land drains. Screens were especially made for this purpose during the early nineteenth century. If required, a wheelbarrow could be used in conjunction with the screen, which would necessitate a framework extended above the body of the wheelbarrow into which the screen was received. When the wheelbarrow was absent, the screen was raised at one end on its own framework, forming an angle of 45 degrees to the ground. The mixed stones were cast by a shovel onto the highest part of the screen and they immediately fell through the wirework that formed the base of the screen, into a heap below. If they were larger than the apertures of the wirework, they tumbled down over the wire into a chute which directed them to a hopper. A second screen, with smaller apertures, could be placed beneath the main one if necessary, in order to screen the smaller stones a second time. As a result, three separate heaps of varying sizes were formed. A similar arrangement is nowadays employed in quarries and by builders etc. where stones need to be graded.

The McAdam pipe layer was an asset when ground was full of stones and gravel and the bed of a draining trench unavoidably rough. In such conditions, an ordinary pattern of pipe layer was not satisfactory and unless pipes were joined together by the use of collars, they were likely to be out of true and consequently block the flow of water. To avoid this difficulty, Mr McAdam of Somerset designed during the 1880s an instrument to hold a line of pipes securely in place on an uneven surface, whilst workmen packed soil around them. It consisted of an oak or ash wood pole about 7 ft. in length and of a diameter just sufficient for a string of drainage pipes to be slipped onto it end to end. The pipes, thus contained along the pole, were laid in the bottom of a trench and after the soil had been

firmly rammed around and above them, the pole was pulled out by means of a handle at one end. The pipes remained in the ground and the drain was completed section by section. This instrument was rather late on the scene, as by the 1880s draining machines were beginning to excite attention, but it remained useful to those laying pipes by hand.

The drainage level. When land drains were under construction it was possible, by means of gauges, to confirm that the bottom of the drain was flat and not likely to retain water in any hollows or depressions. For this examination three identical gauges were necessary. Each took the form of a vertical rod onto which was slotted a brass cross-piece that was made to slide up or down or be maintained by a screw. To use the gauges, the cross-pieces were first placed at identical heights on the rods and the first gauge was then positioned vertically at one end of the drain with its foot resting on the bottom, and the second gauge was placed likewise at the opposite end of the drain. The third gauge was gradually moved along the length of the drain by a man who ensured that its foot was always in contact with the bottom. It was then possible for a second man to sight a line between the brass cross-piece on the two fixed gauges and to observe whether or not the cross-piece of the moving gauge kept the same level. If no variance occurred then the level of the drain was satisfactory. If however it fell below that line, then too much earth had been removed, or if it came up above the line then the bottom of the drain had to be further cut away. Such devices in drain cutting were seldom used before the nineteenth century. Various plumb-levels were available for retaining the correct shape of drain but in most situations the experienced eye of the drain maker was considered to be reliable enough.

The McAdam pipe layer

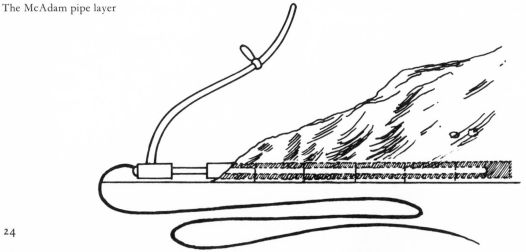

Implements and Machines for Land Drainage

The trenching plough, illustrated by Walter Blith in his *English Improver Improved*, 1653, was probably the first implement designed to cut out trenches for the purpose of land drainage. Its beam contained a single knife-edged coulter which made a vertical incision, about 12 in. deep and cut one side of the trench as it moved along. The second or opposite side of the trench was cut parallel to the first on the return journey, and the earth between the two was then excavated with a gouge or spade. Blith did not

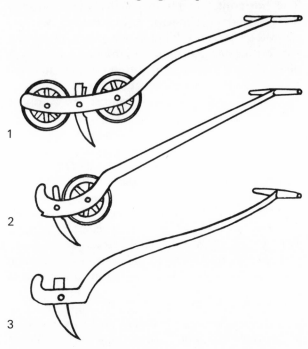

Seventeenth-century draining ploughs from Walter Blith's *English Improver Improved*, 1653:
1 Trenching wheel plough
2 Single wheel plough
3 Plaine trenching plough

disclose any dimensions or the method of drawing the plough, but it is likely that animals were harnessed to the fore-end of its beam whilst the driver walked at the rear holding the single stilt, in which position he would be able to steer the plough and hold the coulter at a constant depth. There appear to have been three different forms (illustrated above), one with a single wheel, one with two wheels and one without wheels. Inventors were soon striving to produce more efficient draining

ploughs, which would cut out both sides and sole of the trench in a single operation. Such implements did emerge early in the next century but generally they were found to be clumsy, ineffective and so heavy as to require an unwieldy number of horses to draw them across wet land. Blith's plough was probably used alongside them for a good many years, as it was again recommended in 1716 by John Mortimer, in his *Whole Art of Husbandry*.

The Cambridgeshire draining plough was at that time requiring the force of twenty horses to excavate a trench 12 in. deep and 18 in. wide. This plough formed the basis of most others over the following years and was probably derived from Blith's plough. It was fitted with two knife coulters, instead of the single one, and they were placed in parallel on either side of the beam, 18 in. apart to cut each side of the trench, whilst the sole was cut by a wide flat share. Turf and soil were thrown clear of the trench by an exceptionally long mouldboard.

Draining ploughs. During the 1760s the Royal Society of Arts awarded prizes of 50 guineas each to Mr Cuthbert Clarke of the Isle of Wight, and Mr Knowles of Belford, Northumberland, for the design of efficient draining ploughs. Clarke's plough was constructed of heavy timbers reinforced with iron, and carried a roller at the fore-end, its function being to prevent the share from penetrating too deeply into the earth. The roller was equipped with three sharp projecting rims to slice through the turf, and behind it were three coulters, and a wide flat share to cut the vertical sides and flat sole of the trench. The loosened soil was raised up and cast a good distance to either side of the trench by two long sweeping mouldboards. Both this plough and Mr Knowles', of which some two dozen were made, were generally considered too heavy for work in clay. A heavy draining plough made by Mr Tweed of Sandon required the force of six horses to cut a trench 12 in. deep and it also laid a length of twisted straw rope in the bottom. Other drain cutting ploughs followed spasmodically, the most notable by Mr Duckett, who also produced a three furrow plough, whilst Grey's draining plough ran on two wheels, one either side of the trench. Almost all of them were based upon the earlier Cambridgeshire draining plough, but none of them seems to have worked any better. Experiments continued into the nineteenth century but by that time enthusiasm had dwindled and various forms of underground drainage were then attracting attention. The **mole plough** and the **subsoil plough,** were soon to be perfected, and then there was no longer any great urgency for a surface draining implement.

Drain boring-irons were first utilised by Joseph Elkington in 1764 to pierce underground springs and direct the pent up water into drains. His system was awarded £1,000 by the Board of Agriculture. For many years afterwards the irons were used by farmers to rid their land of water which had collected there in the form of large puddles or small lakes. To fulfil this objective it was necessary to pierce the impervious strata that lay immediately beneath the water, or, if the water was lying upon clay, to pass down through the clay and to penetrate limestone rock or any other hard substance below. The principle items of the boring set were the oak cross-handle and four iron extension rods which the operator used to screw cutting irons down into the earth. This cross-handle was provided at its exact centre with a socket into which the first iron rod was screwed. All the extension rods could be joined together if necessary, by means of screw joints. This enabled the cutting irons to reach 12 ft. into the ground. On commencement of the drilling, a pyramidal punch was screwed down into the ground to a shallow depth in order to provide a location for an auger. The auger was about 3 in. in diameter and

18 in. in length, with a sharp cutting edge and hollow body which removed a core of soft substance as it descended, leaving a neat, round hole. If rock or any other hard material was encountered, then the auger was removed and a chisel-pointed iron was attached to penetrate it. The irons had to be lifted free of the hole at intervals, and the auger cleaned and inspected. Further extension rods were attached if a greater depth was required. Elkington's boring-irons and his drainage system do not seem to have been much used after the turn of the eighteenth century. Penetration of the earth with boring-irons was made much easier with the advent of portable steam engines during the late nineteenth century.

The horizontal auger invented about 1790 by a Mr Heafield of Hathern, Leicestershire, was acclaimed by Joseph Elkington and was employed for a short time on land drainage in the midland counties of England. It was intended to lessen the time and expense of cutting water courses and performed its work in a neat and precise manner, excavating a sufficient passage for water in which lead pipes could be laid without the necessity of opening a deep trench. It was also used for tapping springs and for finding water at the bottom of a hill, in order to supply a village or drain the land. Unfortunately, it was limited to horizontal boring, and being an expensive machine did not come into general use, but some were made and used whenever the occasion was demanded by contractors. The general arrangement of the auger can be seen in the accompanying illustration. The mechanism was made in brass and iron and was contained inside a flat wooden framework 8 ft. 10 in. in length and 2 ft. 10 in. wide. When it was being used to penetrate a high bank or hillside with the intent of tapping a spring or releasing a volume of pent up water, the level of the bottom of the water had to be estimated and followed to the outside of the hill where the bore was to be made. At that position an area was

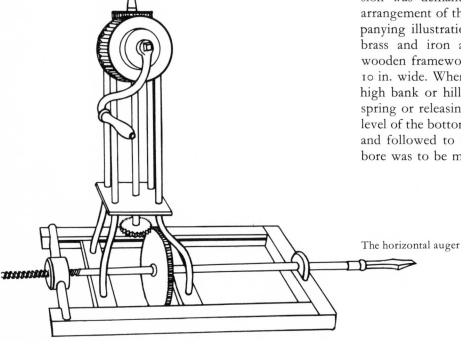

The horizontal auger

cleared so that the frame could lie almost level but with the auger slightly inclined upwards. Two men were required to work the crank handle on top, and toothed wheels transmitted its motion to a horizontal screw shaft which supplied the necessary

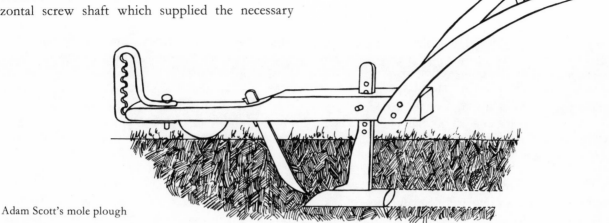

Adam Scott's mole plough

pressure and rotary action to the auger-bit. The auger was made as a hollow shell in order to collect a core of earth as it penetrated the hillside, being drawn back after every 4 ft. of progress so that it could be emptied and another extension rod, 4 ft. in length, included in the line. This was simply accomplished by reversing the handle. Providing the work was not interrupted by rock or stone, two men could accomplish between thirty and forty yards of work in one day. Hard clay presented no difficulty, but stone delayed the work and required a chisel to be attached in place of the auger. The use of such a machine was no doubt hindered by the expense of its construction and it does not appear to have continued in the following century.

The mole plough may well have been invented by a Mr Adam Scott of Essex, as his plough was reviewed by an Agricultural Committee sponsored by the Royal Society of Arts in the winter of 1795–6. In the following year a patent was recorded for Harry Watts but it was claimed that such ploughs, drawn by twenty horses, had existed in Essex for one hundred years or more. Scott claimed and duly exhibited that his plough was capable of forming a continuous horizontal tunnel or drain, 3 in. in diameter, 12 to 18 in. below the surface without turning up any soil. The demonstration took place on the clay land of Marybone Park, Middlesex, and the mole plough, drawn by six powerful horses, managed to complete 200 yards of drain in two days, with the tackle having broken several times.

However, the committee were impressed by the resulting bore which they found to be 'perfectly hollow, circular and as sound as a leaden pipe'. The plough under inspection, consisted of a strong ash beam about 5 ft. in length with a single stilt at its rear end. There was none of the usual apparatus, such as digging share, mouldboard or coulter contained below the beam, only a pointed iron mole 12 in. in length and 3 in. in diameter at its largest end. It was secured to the beam by the use of two flat iron bars, both of which were sharpened along the fore edge and passed up through slots in the beam where they were held fast by pins in order to secure the mole at the intended depth of drain. The inventor modified his plough in 1797, when he omitted one of the two iron bars and replaced it by a knife-edged coulter, which sliced through the soil in advance of the mole and enabled the work to proceed at a faster rate.

Comparison of these early mole ploughs with those of a half century later show little development except for the addition of wheels or rollers beneath the beam to regulate the depth of work. The credit for attaching a wheel to the front of the beam might perhaps go to a Mr Knight of Thaxted, Essex, who manufactured mole ploughs in the early years of the nineteenth century. A gallows, with two wheels, was later added by Mr John Vaisey of Halstead, and Richard Lumbert of Wyck Rissington, Gloucestershire, used two rollers instead of wheels. Strong forces were still required to draw them through clay, and teams of twelve or more horses continued as the main source of power, being sometimes replaced by a windlass and chain which was anchored at one end of the field and pulled the plough towards itself, then moved with the plough as each drain was completed. Not all farmers owned tackle of this kind, due to the high cost of purchase and operation, but there was no shortage of contractors, such as Richard Lumbert or Mr Rogers, also of Gloucestershire, who offered their services throughout the kingdom and completed 200 to 300 perches of drain per day. Such drains were not permanent since they relied so much on the nature of the soil through which they passed, and even if clay they would have collapsed after only a few years' service. Mole drainage remained exceedingly popular for the duration of the nineteenth century and it proved to be quite efficient when a series of parallel moles had been formed at a depth of about 18 in., all connected with a deep main drain to carry the accumulated water away.

Mole ploughs came into more widespread use when steam engines provided a cheap, reliable form of haulage, and in the twentieth century they were re-designed for attachment to powerful track-laying tractors. The essential part of the plough is still the cylindrical mole which is secured to the bottom of a strong iron bar attached to a frame.

Mr Lumbert's mole plough, for which he obtained a patent in 1800, could be purchased at a price of 50 guineas or the inventor would go to any part of the kingdom with his machine and working team, which consisted of eight women, who were paid 8d. each per day to work the windlass, and a foreman who adjusted the plough. He was paid over 7s. per day to supervise the team and was responsible for replacing the long iron link chain whenever it snapped. This plough was similar to other mole ploughs except that it travelled upon a large iron roller at the fore-end of the beam and a smaller one at the rear instead of wheels. There was also an extra sharp-edged coulter before the mole. The first operation in setting the apparatus to work was to position the windlass at one end of the land, in line with and facing the plough at the other end, the distance between them being up to 60 yards. Next, the mole attachment on the lower end of the coulter was set 18 in. below the beam and was dropped into a narrow hole, dug out beneath the plough. The windlass was anchored securely to the ground and after its chain had been passed around

a pulley on the fore-end of the plough, the eight women commenced to wind the chain back onto its cylinders. This caused the plough to move forward whilst the combined weight of the beam and rollers ensured that the mole remained at a constant depth below ground, the narrow sharp-edged coulter to which it was attached sliced vertically through the soil, and below the pointed mole left a cylindrical passage in its wake. The sides of the vertical slit, cut by the coulter, would close together when the rear roller of the plough passed over, leaving the lower cylindrical passage intact. The plough moved forward continually at an average of 5 yards per minute until the full length of the drain was completed, then a deal of time was expended on moving the windlass and plough into position to cut a parallel drain between 3 to 6 ft. away from the first one. In this manner the extent of the land was undermined with passages, each receiving water from the land above and carrying it to a main drain, whereby it was removed to an underground tank and later returned to the land by means of a wind pump or **liquid manure cart.** A Mr Rogers of Withington, Gloucestershire, adapted this heavy plough to work with a capstan and one horse before it was mechanised by John Fowler.

John Fowler's mole plough was exhibited at the 1851 Royal Show, where it astounded agriculturists by the manner in which it laid a string of wooden pipes in the ground at the depth of 4 ft. The only surface digging required was at the start of the pipe laying operation, when a narrow hole was dug out to accommodate the mole at the required depth. The idea was not originally John Fowler's, as some mention, and correspondence appertaining to it had been featured in *The Gardeners Chronicle* some years before. He was however the first to prove its practicability and built his plough as a low four-wheeled carriage with the mole attached below. He then proceeded to try it out on his own farm and on those of his neighbours. It was drawn along by a windlass, operated by a pair of horses who walked round and round to wind in the chain, and pulled the plough towards them. Attached to the rear of the mole, was one end of a long line of wooden draining pipes, all threaded together along the length of a wire rope so that when the plough moved forward with the mole piercing its circular passage, it pulled the string of pipes into the ground behind it. All that remained to be done when the plough had reached the end of its journey was to release the rope and draw it back through the pipes, leaving them buried in the soil. The Royal Agricultural Society's *Journal* of 1851 reported of Mr Fowler's draining plough, 'which by an "invisible" wire rope draws towards itself a low framework, leaving but a trace of a narrow slit on the surface. If you pass, however, to the other side of the field, which the framework has quitted you perceive that it has been dragging after it a string of pipes, which still following the plough's snout, that burrows all the while four feet below ground, twists itself like a gigantic red worm into the earth, so that in a few minutes, when the framework has reached the capstan, the string is withdrawn from the necklace and you are assured that a drain has been invisibly formed under your feet.' In the following years Fowler's apparatus was employed to lay a great number of porous clay pipes in the area of Brentwood, Essex. It was also used at Wormwood Scrubs, but by 1854 the inventor had progressed to using steam power as a means of hauling his mole plough and gained a silver medal for his efforts at the Lincoln Show that same year. Its manufacture was then taken up by Eddington's of Chelmsford. The experience gained with this particular activity quickly led him on to his greatest and revolutionary achievement, that of successful **steam cultivation,** but his company continued to manufacture mole ploughs for many years afterwards.

Pipes for supplying natural gas, water, oil and land drainage are nowadays laid in an almost identical manner by powerful tractors with a mole attached at the rear end. The pipes are, however, made in flexible plastic, to various diameters, and are continuous along their length.

Mr Beart's patent machine for making drain tiles. Until 1840 land drainage tiles of various shapes and sizes were prepared by hand, their cutting, moulding and firing being normal procedure in brickyards at that time. With the increasing desire of farmers to improve their land by drainage, came the necessity for faster methods of tile production. Clay draining pipes appear to have been manufactured in the first place at the end of the eighteenth century by John Reade, who was a gardener at Horsmonden, Kent. He made them by bending a sheet of clay about a wooden shaft some 3 in. in diameter. In 1840 Mr Beart invented his tile-making machine which, although it did not fully mechanise the process immediately, did inspire other inventors over the next three years to construct machines that were capable of performing the whole process with the minimum of supervision. The principles of these machines are used at the present day. Most of the early tileries that catered for the farmer seem to have been located in south-east England. Mr Beart's machine was itself improved by Ransome's in 1843. In its original form, Mr Beart's machine consisted of a rectangular iron chamber mould, 13 in. long, 10 in. wide and 6 in. deep, open on the upper side and mounted upon four legs of convenient height. The prepared clay was placed inside the chamber and rammed down with a mallet to ensure that no air pockets remained. By turning a hand cranked cog wheel and ratchet, the operator pushed up the base of the chamber, and by so doing forced the solid block of clay to ascend with it and gradually to project up beyond the top edges of the chamber. The mechanism was arrested at every 1 in. of the piston's ascent, and at such intervals the operator would cut off the projecting clay with the aid of a **strike.** The operator's assistant then lifted the slab of clay with a spatula and placed it to one side upon a flat wooden board, or if it was to be a horseshoe tile, draped it over a bending block into its final shape. The operation was repeated six times until the chamber was emptied. The assistant who removed the slices, then cleaned out the chamber with a damp rag and scraper, and whilst the operator was engaged upon refilling, he would wash down the bent tiles and remove them to the drying shelves. Although only partly mechanised, this machine with one operator assisted by two boys could manufacture 3,000 tiles per day.

Left: transverse section of Mr Beart's tile-making machine
Right: tile horses or moulds; the strike; the finished drain tile

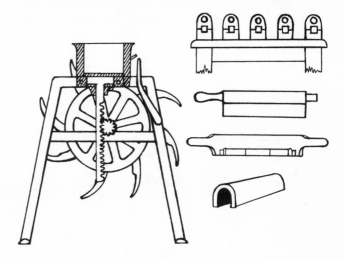

The tile-maker's strike, as used with Mr Beart's machine for cutting clay, was made in hardwood some ¾ in. thick. The tile-maker probably had a selection of strikes close to hand, all of them different in length, but each one convenient for a particular size of tile. His strikes were fashioned from one single piece of wood, having a hand grip on either side and a recess 2 in. in depth cut out of the bottom edge. A thin brass wire was fixed taut across the mouth of the recess and was used by the operator, to cut the clay into slices.

The Marquis of Tweeddale's tile-making machine had increased capacity to 10,000 tiles per day within two years of Mr Beart's invention. Developed by the Marquis for his Patent Tile and Brick Company of London, his machine and production system were largely responsible for further development that resulted in modern brick and tile-making machinery, and machines similar to that of the Marquis of Tweeddale's were used until recent years in many brickworks throughout the world. Clay was shovelled into the machine by workmen and was then thoroughly crushed by the action of rollers and forced by their compression through an aperture onto a moving web of canvas. In its flattened state, the clay was thus conveyed through moulds which gave it the necessary shape required for a tile. After this formation, the endless length of clay was sliced through at required intervals, by a thin taut wire which was mechanically operated and timed to cut each tile to exactly the same length. The tiles were then carried away by another endless web to the drying sheds, where they stayed until a firing kiln was ready. Tile-making machinery has since undergone little further development.

A 'patent horizontal pipe and tile machine' produced by Messrs Armitage and Itter was suitable for contractors and farmers who were inclined to manufacture their own land drainage pipes. This portable cast-iron machine could be operated by a strong boy who turned the crank handle to compress clay and force it through a die of any size up to 5 in. internal diameter. It produced continuous pipes which were sliced into equal sections by the cutting wires hinged onto the side of the long roller topped table. The price of this machine in 1885 was £16 and various size and pattern of die were provided.

The subsoil plough was devised by an ingenious Scotsman, James Smith of Deanston, to break up subsoil without moving its position or turning it into the layer above. It was not until the inventor effectively used this implement to improve the state of his waterlogged land that it was brought into prominence, although a form of subsoil plough, known as 'the miner' had existed in England for many years previously. The miner plough simply consisted of a share without any form of mould-board, attached below a wooden beam. It was drawn by six horses along the furrows left by a common plough where it penetrated to a further depth of 12 in., breaking and stirring the hard packed subsoil. Its use did not become widespread, as an essential accompaniment was thorough land drainage—a necessity overlooked by most farmers at that time, but one that was brought before them most forcefully when James Smith constructed his own subsoil plough and employed it in conjunction with a system of parallel land drains. The results of his attempt to combine subsoil ploughing with underground drainage proved so remarkable that his plough, his method and his farm became famous throughout Scotland and England. In 1823, he had taken over the 189 acre Deanston farm, the land of which was mostly wet and foul with an abundance of rushes. He attempted to improve its condition by laying a series of parallel drains, 2 ft. 6 in. deep, at distances up to 21 ft. apart, by which the water was carried away from the fields into 3 ft. deep receiving drains, and then passed along 4 ft. deep main drains, into underground tanks. The V-shaped drains were filled to half their depth with small stones and covered over with soil. They proved as effective as they had done many times before, but James Smith

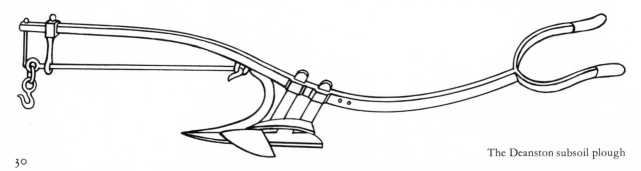

The Deanston subsoil plough

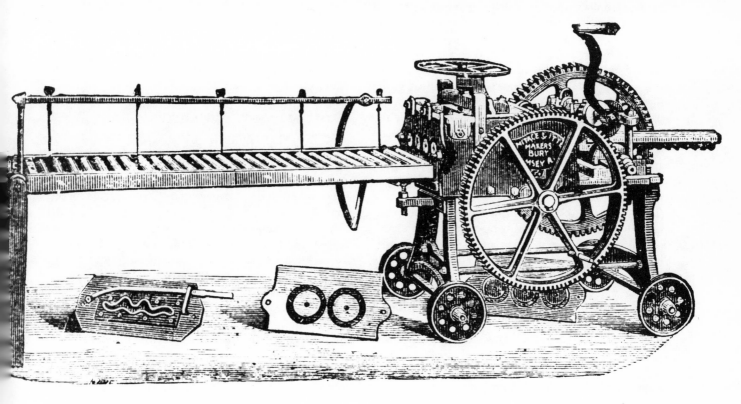

Drainpipe making machine. Armitage and Itter

realised that if the tightly packed subsoil between the drains could be broken up, without being brought to the surface, it would allow much more water to filter into the drains. For this purpose, he constructed his plough and utilised its great weight to penetrate between 16 and 18 in. in depth which shattered the subsoil and allowed air to meliorate it. The astounding results were witnessed by his neighbours and soon his farm was inundated with visitors from different parts of Europe who desired to see the thriving land which was once a marsh. Smith's subsoil plough had an unwieldy length of 15 ft. and was first constructed of huge timbers, then later in iron, to withstand the strain of a six horse team whilst working anything up to 20 in. below the surface of the land. At such times the beam and stilts would have been the only part of the plough visible above the soil whilst below were a curved coulter and a pointed share. An iron wing was attached to the rear of the share on the furrow side and, as the plough was drawn forward, the subsoil, penetrated by the points of the share and coulter, raised up and crumbled as the wing passed

through it. Smith's plough gave excellent results, especially when it was preceded by a common plough cutting the furrow. The final operation was to remove, by hand and crowbar, any rocks or buried obstacles which had hindered the plough. The separate operations of subsoiling and opening the furrow involved the landowner in extra expense with regard to both time and labour, a matter which prompted inventors to consider the possibility of an implement that would condense the separate actions and reduce the overall expense of subsoil ploughing. Amongst others, Sir Edward Strachey, of Rackheath Hall, Norfolk, appears to have produced a very practical version of the subsoil plough during the 1830s.

The Charlbury subsoil plough was the first of many ploughs which stirred the subsoil and also opened the furrow along which it was working. Although it did not stir the subsoil as thoroughly as Smith's implement, it did set a pattern in subsoil ploughs for the remainder of the nineteenth century. It consisted of an ordinary plough framework with a single subsoiling tine at the rear end of the beam, near to where it joined the stilts. This tine resembled a huge coulter, being curved forward so that its broad sharpened edge was immediately below the sole of the plough. It was adjustable in its working depth between 12 and 18 in. and effectively stirred the subsoil whilst the furrow was opened by a share and coulter attached in their common position to the plough frame. The depth of the furrow was regulated by a single wheel at the fore-end of the beam.

Many different subsoil ploughs were produced over the remainder of the century, and subsoiling attachments were provided for fastening to other forms of plough. A notable implement employed for subsoiling after 1870 was Ransome's patent double plough; elegant and low lying, it was made entirely in iron with the subsoiling tine set in front of the coulter. Steam power provided the opportunity for larger subsoil ploughs. They were mostly constructed in the form of experiments, as the subsoiling work could readily be achieved in an uncomplicated manner by a **steam balance plough** equipped with subsoil tines. Where steam haulage was not available, horse teams continued the hard work of subsoiling, but when the tractor came into use, subsoil ploughs were designed especially for attachment to it. They were little more than a number of tines, similar in shape to that on the original Charlbury plough, arranged across a bar, or alternatively some farmers removed the first body of a three furrow tractor plough and attached a single subsoil tine in its place.

The drain cutting wheel. In their 1801 *Account of the Mode of Drainage,* the Board of Agriculture mentioned an invention called the draining wheel, which according to their report was regularly completing twelve acres of drain cutting per day in the county of Essex. The Board were somewhat vague about its working action and manner of construction, but it would seem to have been a spoked wheel, cast in iron to a diameter of 4 ft. with a rim 4 in. in width

and $\frac{1}{2}$ in. in thickness. It carried a number of sharp, radial tines placed at regular intervals around the circumference for the purpose of digging, and the whole wheel was arranged inside a wooden framework, weighted to determine its working depth and drawn along either by horses or by a windlass and chain. The Board did not supply any details regarding its digging action, but it is probable that the wheel was made to revolve quite rapidly which caused the tines to cut into the soil and throw it back for some distance. The resulting trenches were $\frac{1}{2}$ in. wide at the bottom, 4 in. in width across the top and 15 in. in depth, but could be varied to suit the soil conditions by the amount of weights placed in the framework. They were filled with small bushes, bundles of twigs, twisted straw-rope or stones and covered over with light porous soil, or they were left uncovered in order to be split wider and deeper by the heat of the sun. The wheel seems to have worked best when the land was wet, but despite the Board's enthusiasm it was not again recorded.

Paul's rotary drain cutter appeared in the 1850s. It was invented by a Norfolk gentleman, Mr Paul of Thorpe Abbots, and designed to ease the labour of hand trenching. But as with the earlier drain cutting wheel, its success, if any, appears to have been short-lived. It comprised a heavy framework which lay flat upon the ground and contained a vertical wheel 4 ft. in diameter with fifteen digging blades set around its perimeter. A hole was excavated at the commencement of work and the wheel, set to dig the required depth of drain, was hauled by windlass and horses. The winding chain was attached by one end to the windlass and passed around the wheel, where its links located onto short iron teeth. As the chain was wound in it caused the wheel to revolve, so that the digging blades came up against the soil and cut it away as a continuous trench. Neither Mr Paul's wheel nor the one before it had the necessary

Paul's rotary drain cutter.
From J. C. Morton's
*Cyclopedia of Agriculture, c.*1856

Robson-Hardman drain cutting machine

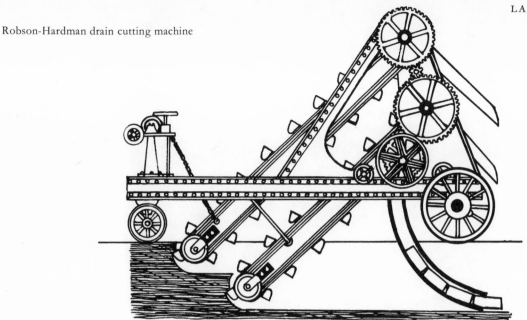

provision for laying excavated soil to one side, and most likely it was carried over the wheel by the blades and thrown back into the trench behind. Rotary drain cutters were not brought to perfection until the present century when powerful steam and oil engines were employed. The American company, Holts, produced a huge steam-driven crawler-tractor with a drain cutting and automatic pipe laying device at the rear. Some were imported into Europe during the 1920s.

A drain cutting machine made by Messrs Robson and Hardman was exhibited at the Derby Royal Show in 1881 and represented the first of a series of attempts, by those manufacturers, to successfully mechanise the process of opening land and laying pipes. Other attempts had been made earlier in Great Britain and America. Some success was achieved by A. and W. Eddington of Chelmsford, who adapted John Fowler's drain plough to that purpose. The Robson Hardman machine utilised an endless chain of shovels or buckets which excavated the soil whilst at the same time it laid a line of cylindrical clay pipes and replaced the soil over them as it moved along. The whole machine was about 12 ft. in length and made wholly of iron, with two large wheels at the rear for travelling and two smaller wheels at the fore-end for steerage. It was hauled along on the end of a cable in similar manner to a steam plough, and required the direct attention of only one man to feed in a supply of pipes. The digging shovels were chained together in the form of endless elevators and two such sections were employed, mounted in tandem, with the rear one digging deeper in order to remove the subsoil separately and then return it to earth before the top soil. The elevators worked in identical manner and both derived their motion from the large wheels at the rear end of the machine. When the shovels reached the full height of their elevation, the soil was tipped into chutes and directed over the tail end of the machine, where it was deposited on top of the pipes which had descended through a curved conductor into the newly opened trench. Provision, in the form of a winch and chain, was made for raising and lowering the elevators to suit the required depth of drain. The Royal Agricultural Society judges did not award the manufacturers a prize medal at the 1881 show, since the machine was not put through sufficient trials to prove its capabilities, but they were of the opinion that it did not fully overcome the difficulty of laying pipes accurately in contact, end to end, neither did it provide a proper and equable fall along the length of the drain. However, the principle was recorded as showing some promise but did not become fully practical until powerful diesel motors were available in the twentieth century. Some modern machines excavate trenches in exactly the same manner as the 1881 machine and some by rotating discs which throw the soil clear of the work, whilst others employ an Archimedal screw-cutter. The majority of trenching is nowadays completed by mechanical scoops or shovels.

33

2

Cultivating the soil

The Plough

Historians are generally agreed that the primeval implement must have been a straight, pointed stick, first used to dig up earth during the search for edible bulbs and roots, and then later made much larger and perhaps fashioned with a short projecting side piece, so that the user could exert the pressure of his foot to gain deeper penetration.

Early man's struggle for survival became closely bound to the success of his crops and so he was forced to cultivate a greater area of land in order to grow more corn and ensure a harvest large enough to sustain his life. As always the crop was better for a thorough preparation of soil, the whole of which was carried out by hand labour, using in the first place the simple digging stick, then later as his brain developed, a crude form of hoe or hoe-plough. It was little more than a forked branch with one limb cut short and pointed to scratch the earth whilst it was hauled along by a man who grasped the longest limb.

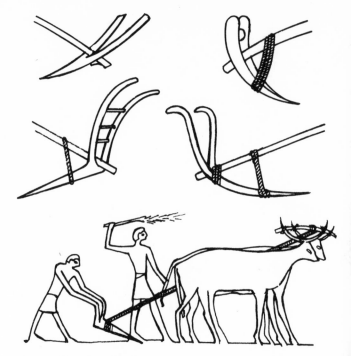

Ancient Egyptian ploughs

Ancient Egyptian hoes

Considerable advance was made by the ancient Egyptian farmers of the Nile valley. They succeeded in growing flax, millet, wheat and barley, and reaped the harvest with flint sickles. In their dry climate, controlled irrigation was necessary for growth and so to secure the effects of an annual flooding, they built dykes, excavated canals, and raised water to inland regions with the assistance of water wheels. Perhaps it was here that the digging stick evolved into the type of hand hoe often found sculptured on ancient Egyptian monuments. It was, as illustrated, two wooden parts, one of which was slightly curved

and sharpened at the end to form a blade for digging, whilst the other piece was a straight wooden handle fastened to it by thonging. The end of the blade was shod with flint or copper, and the instrument was used with a chopping action to break down large clods of soil and to create a tilth suitable for seed sowing. The seeds were then covered against the birds by use of a sharp pointed **bush harrow** or by driving a herd of cattle up and down the field. The changing of hoe into plough must have been in very very early times, as a seal of about 3500 B.C. shows the Egyptian plough to be already well formed. It was fitted with a triangular iron share, but even so was only capable of opening a shallow furrow by pushing the soil away to either side. Primitive ploughs were all the same in that they did not undercut or turn over the soil. The Babylonian plough of about 1300 B.C. (illustrated on page 108) had an iron-tipped share, and a bamboo tube attached to the side through which seeds were dropped into the furrow as the plough moved along. But it seems to have been an oddity and did not come into general use. In the first place, oxen were tied directly to the plough handle by their horns, but a crude form of yoke was soon evolved by lashing a beam across the horns of a pair of oxen.

The plough was important in Greek agriculture and came to them from the Egyptians along with

Chinese plough, from the *Farmer's Magazine*, 1805

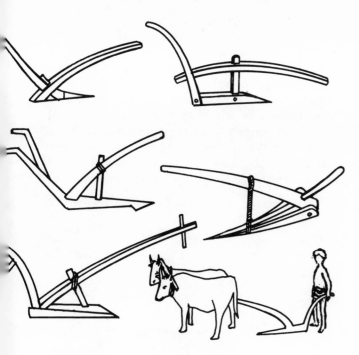

Primitive ploughs

other implements, tools and customs. Theirs was a light manœuvrable plough built wholly of timber except for its iron share. Hesiod described it as consisting of three parts: the share beam in oak, the draught beam of elm and the plough tail, all secured together by pegs. Virgil called it the crooked plough because the beam was curved forwards to the draught animals. A ploughman walking behind caught hold of the upright tail piece, and controlled the depth of ploughing by throwing his weight onto the rear end of the share beam.

The Romans used a wide variety of ploughs and also other cultivation implements, all of which were adapted to different districts, climates and types of soil. They had a variety of **harrows, rakes, spades, forks, mattocks** and **hoes.** Cato mentioned two ploughs, the *Romanicum* for stiff heavy land, and the *Companicum* used on lighter soil. Varro described a plough with two wings like small mouldboards attached to the share for the purpose of covering over sown seeds. Pliny mentioned a plough used for the same purpose, whilst he described the invention of a wheeled plough to the inhabitants of Rhaetia in the first century A.D. They sometimes equipped their plough with a knife coulter which cut the land open before the oncoming share.

Some primitive ploughs were used in different parts of the world until quite recently. They were

of a very crude form, being little more than scratching implements, no further advanced than those used by the ancient Egyptians and Greeks. In the Near East, India, South America, and parts of Europe, notably Sicily, Spain and Portugal, the plough was exceptionally crude and simply tore open a shallow rut as there was no provision for undercutting or turning over the soil. All primitives had the same type of plough, a selection of which are illustrated.

The Celtic plough of the pre-Roman years was very primitive and only made a shallow rut or furrow, casting the soil slightly to either side. The narrow strips of unploughed land left between the furrows was a hindrance to the Celtic farmers, but to overcome this they cross-ploughed their land, that is, ploughed it twice, the second time with furrows at right angles to the first. This ensured that all of the land was ploughed and the work was completed most conveniently where all sides of the field were of equal length.

The Romans invaded and established large farms in Britain. They used an imported two-oxen plough, the *Ard* or *Aratrum*, which was without wheels and suited the light soil, but the British peasants continued to use their former plough and with a team of oxen they drew a furrow 220 yards long, before the plough was turned about. The Roman standard furrow was 120 ft.

The Anglo-Saxons replaced the Romans in Britain, and what little is known of farming over the next 600 years indicates large farms with the land divided between arable, meadow, pasture and plantation, in most cases under a sole owner, but farmed by many men who worked together in a co-operative system. The big fields were divided into many strips, each about 22 yards wide, and they used heavy eight-oxen ploughs to form high ridges for corn and deep furrows for drainage. This heavy plough was featured in Strutt's *Anglo-Saxon Rarities of the Eighth Century* as the Saxon wheel

37

Sixteenth-century ploughman. From an anonymous engraving

plough, and in the seventeenth century was illustrated in Walter Blith's *English Improver Improved*. By that time it was widely used in the midland area of England, where it was known as the Hertfordshire wheel plough. A lighter plough without wheels was used in other parts of England. The changeover from oxen to horse teams was very gradual, and farmers first of all used a single horse at the head of their ox teams for steerage.

Various patterns of plough were used in Britain during the Middle Ages and manuscripts portray farmers using both light and heavy ploughs, with and without wheels, and with sundry patterns of coulter and mouldboard. They were all constructed of timber with digging parts of iron and the mould-board was covered with iron plate. The majority were so heavy as still to require eight or more oxen to draw them through the soil, but in some districts they had been greatly improved, being more suited to the various types of terrain and made light enough to go with two yoke of oxen with one horse in front.

Some of the ploughs had wheels, particularly the Hertfordshire, the Norfolk and the Kentish plough,

38

whilst in other counties they were without wheels. This was the most obvious distinction between ploughs almost up to the nineteenth century, the former being known as **wheel ploughs** and the latter, **swing ploughs.** The size and shape of ploughs varied considerably from one district to another, and even within districts, to about the end of the seventeenth century. So far no rules or guidance had been laid down for their making. Blacksmiths, carpenters and wheelwrights were largely responsible for plough construction and to

Seventeenth-century ploughs from Walter Blith's *English Improver Improved, 1653*
Top to bottom: Plain plough; double plough; Hertfordshire wheel plough; single wheel plough

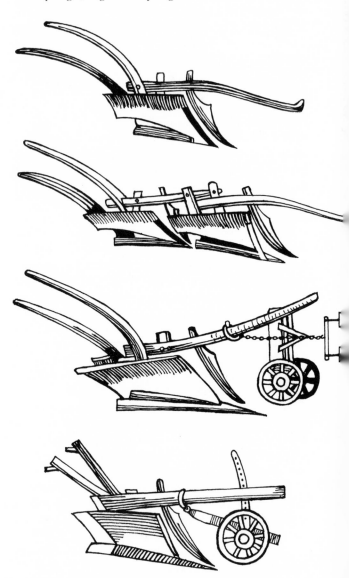

an extent for the design also, as they fashioned plough parts to a pattern already formed by tradition or to the prevailing taste of their neighbourhood. This was the situation at the middle of the seventeenth century when Walter Blith published his *English Improver Improved* and in it laid down some essentials towards improved plough design and construction. He remarked that the then existing timber and iron ploughs were too heavy, the mouldboards too straight, whilst the stilts were generally too short or long and did not afford the user complete control over the implement. These faults, he concluded, were the result of bringing together a carpenter to make plough bodies, and a blacksmith to make plough irons, neither of them having had the experience of using a finished implement. Some of the ploughs with which Blith was familiar and had most probably used are illustrated. Amongst them was the Hertfordshire wheel plough, a powerful implement with a beam 6 ft. long and wheels 20 in. in diameter. Its faults were numerous and it was too heavy for most work other than breaking up stony fallow land, but it continued to be a favourite with farmers in the midland counties. Blith also listed a double-furrow plough, then used in Norfolk for turning two furrows at one time. He described their mode of operation and succeeded in laying down a strong foundation upon which future theorists and designers could elaborate. But things did not change immediately. Some small progress was made over the following half century, but nothing of lasting value. It would seem that inventors were perplexed: they paid attention elsewhere and ploughs failed to find the perfection for which Blith had hoped.

Indeed, the situation had not altered much some sixty-three years later when, in 1716, John Mortimer in his *Whole Art of Husbandry* declared: 'There is a great difference in most places about the make and shape of their ploughs, some differing in the length and shape of their beams, some in the shares, some in the coulters and in the handles etc. Every place being almost wedded to their particular fashion, without any particular regard to the goodness and convenience or usefulness of the sort they are.' Mortimer found a light two-wheeled plough at Colchester and a single-wheel plough in Sussex. The latter he disliked, but the dray plough, common to many districts, he thought to be the best for winter, 'but the worst in summer when the land is hard, because its point is always flying out of the ground'.

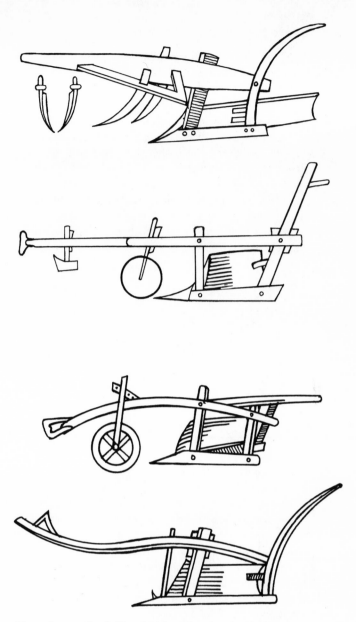

Top to bottom: Cambridge draining plough; Lincolnshire plough; Sussex plough; dray plough. From John Mortimer's *The Whole Art of Husbandry*, 1716

He described the Cambridgeshire draining plough, as used near Caxton, and the Lincolnshire plough, particularly suited to the marshy and fen lands that were free from stones. Its coulter was the sharp-edged Dutch wheel which revolved and cut through grass and roots, whilst a very sharp broad share cut the furrow. Mortimer observed that the mouldboard of the East Anglian plough was made in iron 'by which they make it rounding, which helps to turn the earth or turf much better than any other sort of plough'. His may well be the first reference to a twisted iron mouldboard.

39

About 1730, a plough was brought into use near Rotherham, Yorkshire. For this reason it was commonly called the Rotherham plough, although its origin is somewhat obscure. Various names have been linked with the introduction, but it is generally thought to have been designed and built by one Joseph Foljambe at Rotherham, or to have been brought over by him from Holland. This wooden swing plough with an iron share and coulter and a mouldboard covered with iron plate was considered by all who saw it at work to be more effective and lighter in draught than any other plough in use at that time. A large number of them were made and

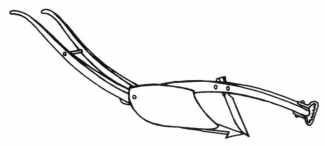

The Rotherham plough

used in the north of England, where they remained exceedingly popular for half a century or more. They were also much appreciated in Scotland, where they were introduced by a Mr Lomax and called Dutch ploughs. But tradition died hard in Scotland and their old plough, drawn by four or six oxen with two horses in front, continued in many districts. It was so inefficient that the land had first to be opened by a special implement known as the **ristle.** The Rotherham plough, however, enjoyed a general popularity, marked the beginning of factory-plough production and its design was taken as far afield as America.

It was an ingenious Scotsman, James Small of Berwickshire, who first applied mathematical principles to the design and curve of the mouldboard. He was born about 1740 and became familiar with many facets of agriculture during boyhood. As a youth he was first apprenticed to a country carpenter who also made ploughs, and then he went to Doncaster to work for a manufacturer of carriages and waggons. In 1763 he returned to Scotland, settled at Blackadder Mount in Berwickshire and engaged in farming. This gave him the opportunity to experiment and put under trial the various ideas he had in mind, one of which was to replace the countless patterns of mouldboard then in use with a

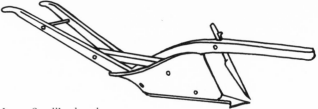

James Small's plough

single, universal shape, which would turn the furrow slice more effectively whilst demanding less labour from the ploughman and the animals. He was seeking the perfectly formed mouldboard, not one curved or twisted haphazardly like the others which caused the furrow slice to topple over through sheer length only. He used mathematical calculations and practical devices to determine the natural twist of the furrow slice as it came back from the share onto the mouldboard, then he reproduced the form of the twist on the surface of a soft wooden mouldboard. He attached this prototype to a working plough frame so that it could be hauled across the land by horses and further observations be made with regard to wear caused by the friction and pressure of soil. When he was finally satisfied with the new shape, it was cast in iron and brought together with other improved parts to form a plough which required less labour and effort from man and beast than the existing unscientific ploughs. The *Quarterly Journal of Agriculture* for 1832 declared: 'By him an implement was constructed which has materially diminished the expense of cultivation, which will answer in every soil, which will turn out the cleanest and deepest furrow with the least force of draught and which, on the whole, is better adapted, for general purposes than any other plough that has hither been seen or heard of . . .' His improved plough was quickly adopted into the improved districts of Scotland, a factory was established to make them in Ireland, and it was gradually accepted into England, where some doubtful theorists tried to claim the design as their own. Despite the jealousy and criticism of others, his plough reigned supreme for many years and beat, in competition, the old Scots plough and the best provided by English manufacturers.

Theory and debate, not only with regard to plough design but to agricultural methods in general, burst forth in a lively manner about the middle of the eighteenth century. It was not to decrease for a hundred years or thereabouts, as the same questions were taken up and furthered by writers in new farming journals and magazines. Opinions were exchanged through the offices of these publications and numerous volumes on the subject of farm management were completed. Perhaps the most effective of all were the agricultural trials and competitions organised by the Royal Society of Arts and others, in which inventors and manufacturers were caused to compete against one another under similar if not identical conditions. In 1767 the Society gave three awards of £50 each for new plough inventions, one of them going to Mr Duckett of Esher, Surrey, for his plough with a newly invented **skim coulter** attachment, designed to bury turf beneath the furrow slice. It was secured through a slot in the beam by wedges and pared off a suitable depth of turf or other top growth which it turned over, before it was buried by the common share of the plough. After only one ploughing the field looked remarkably tidy and showed no evidence of the weeds and grass buried beneath the furrow slices. The skim coulter later became an important feature of 'Match ploughs' used in competitions where cleanliness of work was essential.

The Royal Society of Arts, founded in 1754 did an enormous amount towards the advancement of agriculture and purposely organised their trials in various areas of the country during different seasons of the year, so that competing implements, be they ploughs or whatever, met with a variety of soils and crops. Rivalry between competitors was intensified not only by the considerable money prizes and the medals which were offered, but because the publicity to be gained for the winners was invaluable at a time when manufacturers and merchants were establishing up and down the country. The trials went a long way towards completing a scientific enquiry into plough design, especially when Mr Samuel Moore, Secretary to the Society, invented the **dynamometer,** first used at the Society's 1784 trials, in order to determine the amount of draught required for moving various types of plough and other machinery.

A more robust but equally important manner of contest had become popular by that time. They were ploughing matches, organised on a local or county basis, in which one ploughman matched his skill against another. The first match to be recorded took place at Odiham, Hampshire, in 1784, though doubtless there were others before that. Such matches became popular as public spectacles or as a means of settling wagers and although they were

accompanied by a good deal of carnival atmosphere, rivalry was intense and competitors did their best to complete the furrows with precision. To some degree the merits of one type of plough was determined over another in that manner, even if the duration of work was only a few hours. The real and long lasting effect of such matches was to raise the standard of ploughing throughout the country and to encourage ploughmen to strive for perfection whilst scientific experimentation was left with

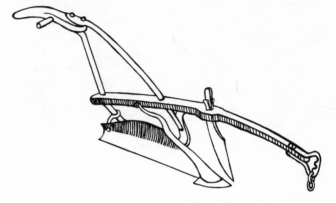

Brand's iron plough, later known as the Suffolk iron plough

the Royal Society of Arts and later the Royal Agricultural Society of England. A notable new plough, made completely of iron and with only one handle, was brought out by John Brand of Essex about 1770. It was an iron swing plough for use with two horses. It cut an exceptional furrow and retained the strength of iron whilst not being too heavy. Brand's one-handled plough, later known as the Suffolk iron plough, gathered an excellent reputation whilst other inventors came forward with iron ploughs; amongst them was Robert Ransome, founder member of the Ipswich firm that still bears his name. His success began in 1785 with a patent for tempering cast-iron plough shares, and his business rapidly expanded from a small concern employing one workman to larger scale foundry production with merchants and stockists in most parts of the country. Further success came to him in 1803, when he developed the self-sharpening chilled share. By cooling the under surface of the cast share more quickly than the upper side, it became harder and wore away at a slower rate, thus providing a permanently sharp edge.

The idea of making a plough which could easily be dismantled and new parts bolted on was patented by Robert Ransome in 1808. The practice caught on

41

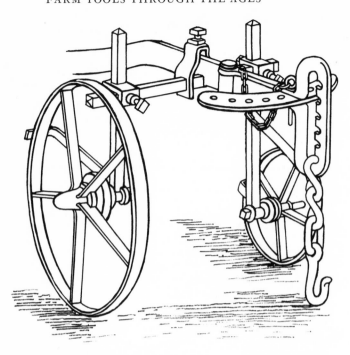

A Ransome plough, showing wheels attached to the beam

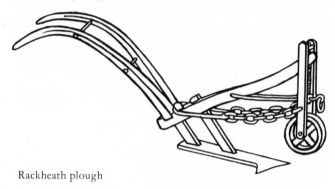

Rackheath plough

with other manufacturers and by the middle of that century a number of universal plough bodies were available. Whilst they were particularly used for ploughing in the ordinary manner, they were later designed to accommodate various attachments such as subsoiling tines, hoeing tines, seeding boxes and potato graips etc., which were used for only a short time and then removed. These adaptable ploughs were popular with the smaller farmers since it saved them buying a range of separate implements.

By 1840, Ransome's factory had no less than eighty-six different types of plough for customers in various places to choose from. A large quantity were shipped to Europe and America, where plough factories, mostly enlargements of the blacksmith's or ploughwright's shop, were only just becoming established. In England, merchants opened their shops in the main market towns and in London where it was possible for farmers to purchase various forms of plough with iron mouldboards or factory-made harrows, hoes and rakes etc. But in spite of this new activity, the use of the new cast-iron implements did not extend much beyond the improved districts. Farmers continued for many years, to use their old familiar implements.

The natural interest aroused by land drainage improvements brought forth the new **subsoil plough,** credit for the invention going to James

Smith, whose enterprising activities, his farm at Deanston and his plough are described in the preceeding chapter on land drainage. His successful methods and his new implements were copied, and various patterns of improved subsoil plough quickly appeared in the field. Amongst the best known was the Rackheath subsoil plough presented in 1838 by Sir Edward Stracey, a gentleman farmer of Norwich, and manufactured in number by Barnard and Joy, implement makers of that same city. It was not only a subsoil plough, as it could be fitted with a device

for raising a whole field of potatoes in a short time, a back-breaking activity previously performed by an army of labourers. Farmers who could not afford to purchase such a specialised plough attached one of the new **branders** or **graips** to the common plough.

The list of ploughs continued to grow, and during the second half of the nineteenth century the number of manufacturers engaged in the production of ploughs and the different patterns available for the farmer were quite considerable.

Plough Parts

The beam was a strong bar of wood or cast-iron, being more or less curved to suit a particular type of plough. It was provided with the necessary locations to accommodate the bridle and the various coulters and was itself secured to the plough frame by two or three bolts.

The frame or body was in the centre position of the implement. It carried a mouldboard attached by couplings with the share fitted to its fore-end.

The slade was the sole of the plough and was bolted to the underside of the body, where it served to support the plough whilst sliding along an unploughed land at the side of the furrow.

The land-cap was attached to the left, or land side of the plough frame and served as a shield against loose soil and stones.

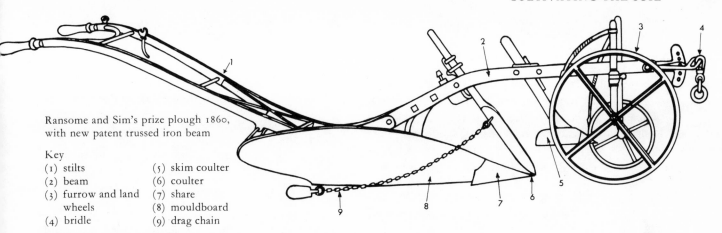

Ransome and Sim's prize plough 1860, with new patent trussed iron beam

Key
(1) stilts	(5) skim coulter
(2) beam	(6) coulter
(3) furrow and land wheels	(7) share
(4) bridle	(8) mouldboard
	(9) drag chain

The mouldboard or breast was made in cast-iron or steel and formed a continuation of the share, though the two were separate items. Its twisted surface was designed to turn over the furrow slice with ease, and lay it close alongside the preceding one. It was attached to the right side of the body frame and was often strengthened by a stay at its rear end to the right hand stilt. There have been many different forms of mouldboard, each one designed for a particular type of work or soil.

The share, a triangular piece of iron with a sharp edge called a feather, made a horizontal cut beneath the furrow slice, slightly preceding the vertical cut of the coulter. It was established immediately before and in line with the mouldboard and its pitch could be adjusted through a lever located in the plough frame, as the work demanded. Different forms of share could be attached to a plough in accordance with the condition of the soil, the difference between them being in the length and angle of the feather.

The stilts or handles were formed separately in iron and curved down to be fixed one on either side of the beam by two or three bolts. They were kept rigid and apart by the use of straight or diagonal braces midway along their length, and their ends were sheathed in wood to provide a convenient grip for the ploughman. Their position was such as afforded him the greatest possible control over the implement, enabling him to maintain a regular depth and correct course for the duration of work.

The drag chain was attached to the rear edge of the coulter and was supported by a second chain from the wheel axle, or it could be attached to the coulter only. It was about 3 ft. in length and had an iron weight attached to its loose end. This was not essential to the plough, but as with the skim coulter

it served to prevent any top growth from marring the appearance of the newly ploughed land. Its position allowed it to drag against the furrow slice as it was turning over and so remove any growth from the edge, causing it to be fully covered by the slice.

The bridle, sometimes called the muzzle or hake, was attached to the front end of the beam and carried the chain by which the plough was attached to the **whippletree.** In order to set or adjust the draught, the bridle was made to swivel horizontally upon the bolt by which it was fixed to the beam, its position being made secure by the small pin which passed through the bridle and the selected hole in the arc-head. This adjustment would alter the 'landing' of the plough or its tendency to move to either side out of line of the draught. The position of the chain upon the notches of the bridle would vary the 'earthing' or working level of the plough, causing it to come up or down as required.

The coulter varied in its length to suit the type of plough with which it was used, and also the angle at which it was attached to the beam differed in accordance with the conditions of work. It was usually positioned so that its point was immediately before the share, its function being to slice vertically through the soil with its cutting edge. The manner in which the coulter was attached was by passing its neck through two loops of a clip that was bolted to the beam. This type was known as the knife coulter. A different form of coulter was fitted to the seed drill.

The skim coulter was attached to many English ploughs and served as a plough in miniature. Its position on the beam was a short distance to the front of the coulter where it pared away any growth on top of the furrow slice before it was turned over.

Although it was not essential it helped towards the tidy appearance of newly ploughed land and was an asset on competition or 'match-ploughs'.

The disc coulter, probably brought from Holland to England during the sixteenth or seventeenth century, was known for a long time as the Dutch wheel. It was an iron disc about 1 in. thick at the centre, tapering down to a sharp-edged circumference. Set before the share point, it was well adapted to cutting through turf and long manure and was often used in addition to a knife coulter, being attached to the same plough beam, preceding the latter and cutting the same line. Its diameter varied a great deal before the twentieth century, but whatever its size it was always set just deep enough for the cut to be made by the underside of the disc rather than the fore edge. Disc coulters are an important feature of some modern tractor ploughs, making the same vertical cut along the furrow wall as the old knife coulter, but with greater effect.

The wheels of the plough were set some distance before the skim coulter and both were adjoined by their independent axles to vertical stems, which were secured on either side of the beam. The wheels were of different dimension, the larger on the furrow side termed the furrow wheel with a smaller wheel on the opposite side termed the land wheel.

The furrow wheel controlled the width of the furrow slice. The axle upon which this wheel had its bearing could be moved horizontally through the vertical stem by which it was held secured to the beam, so that the chosen distance between this wheel and the plough share determined the width of the furrow slice. Vertical adjustment was made possible by the variance of the wheel stem through its eye bolt fixture on the beam. Normally this wheel was placed in advance of the other.

The land wheel was arranged to run on the un-turned land at the side of the plough and its height regulated the depth of the furrow. It was fixed, as the other wheel, by an eye bolt to its side of the beam.

The wheels were scraped clean of mud or soil that would otherwise adhere and lift the plough out of its setting. This was effected by a bowed iron rod, one end of which was attached to the top of the vertical wheel stem, whilst the other end, flattened to form a scraper, was in constant friction with the periphery of the wheel. When wheels were not used, the working depth of the plough was controlled through adjustment of the hitch at the fore-end of the beam.

Some Plough Types and Attachments

Types of plough could be classified as follows:

swing ploughs	ridging ploughs
wheel ploughs	sulky or gang ploughs
double furrow ploughs	potato lifting ploughs
multiple ploughs	steam balance ploughs
turn-wrest ploughs	subsoil ploughs
paring ploughs	draining ploughs

The swing plough was made without wheels and required the attention of a very skilled ploughman in order to keep it well balanced. It would work in less favourable conditions than the wheel plough and was especially suited to heavy or sticky land where wheels were liable to clog. It was also particularly employed in rocky districts where the soil was shallow and the depth of ploughing inconsistent. The swing plough was generally considered to be less perfect than the wheel plough, since its increased draught quickly fatigued the horses and required extra diligence on behalf of the ploughman in order to obtain a uniform depth and width of furrow. However, in some districts of England and more especially in Scotland where it had long been in use, its disadvantages were not sufficient to warrant the use of a wheeled plough and it remained common until about the middle of the nineteenth century. It was finally improved beyond the form devised by James Small, being constructed in iron and adapted more completely to the different soils and circumstances. Amongst the best known were the Argyleshire swing plough, Wilke's, Somerville's, Clarke's, Grey's, Cunningham's and **Finlayson's.** The last inventor was also responsible for a crane neck or self-cleaning plough and later a pattern of self-rid harrow. He also designed a turn-wrest plough.

The wheel plough. During the nineteenth century, manufacturers devoted a great deal of attention to improving the common wheel plough and consequently made it light in draught and easy to handle. One of the ploughs, brought out by Ransome, Sims and Head for the Newcastle 'Royal' meeting in 1864, represents the horse-drawn wheel plough in its most advanced form. Two wheels were provided near the fore-end of the beam, the larger to travel in the furrow and the smaller on the land. The width of furrow was determined by the

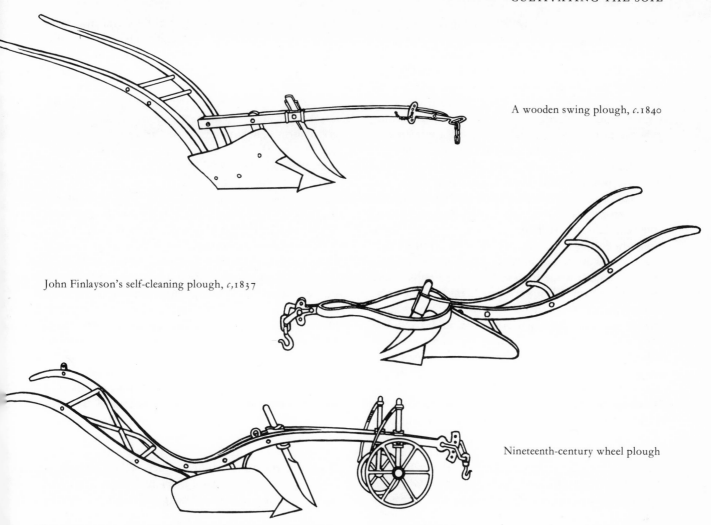

A wooden swing plough, c.1840

John Finlayson's self-cleaning plough, c.1837

Nineteenth-century wheel plough

distance between the furrow wheel and the coulter, whilst the depth of ploughing was regulated by the setting of the land wheel. Both wheels were attached by means of vertical stems and movable sockets to a cross bar that passed under the beam, so making the adjustment of the wheels a simple task. The share and coulter were made with hardened edges, and different mouldboards could be fitted for deep or shallow work. The illustrated plough has a short mouldboard, especially designed for general work on the farm. The stilts were braced diagonally, with iron rods and secured to the beam by bolts, the plough frame being attached under the beam in like manner. This plough was made to different weights and sizes and could be used when the wheels were completely removed as a swing plough. Occasionally the wheel plough was furnished with a single wheel only, and with it set directly beneath the

beam was most adaptable to sloping, 'laid-up' land, where it could be maintained at a constant depth, remaining unaffected by such conditions as would cause a plough with two wheels to work unevenly. It also performed well on sticky land with the single wheel set well out on the land side where it stabilised the plough without hindering the work. If conditions made a single wheel impractical, it could be effectively replaced by a sliding sole. The Beaverstone plough went with one single wheel and was noted for lightness and compact design. It was much used in the west of England during the first quarter of the nineteenth century.

Hornsby's of Grantham, Bushby's of Bedale, Howard's of Bedford and Ransome's of Ipswich gave a great deal of their attention to the design of wheel ploughs and a considerable number of them were in use in England by the end of the nineteenth century. Howard's plough was thought to be the best general purpose wheel plough at the Royal Agricultural Society's Newcastle meeting in 1868.

By the middle of the nineteenth century the wheel plough was of two distinct types. First the Lea or Longplate plough with its gradually curving, convex mouldboard, which was used in autumn ploughing to give a clean, precisely turned furrow, with all grass, weeds, and debris buried beneath.

Secondly, the digging plough, with its short concave mouldboard which was used in the spring to invert the land and reduce it as with a spade. The deep furrow was completely pulverised as it passed over this mouldboard and was left in a light condition under the effects of the weather until spring, when it was easily transformed by the use of harrows into a suitable tilth for seed sowing. Subsequent ploughings were then very much easier. The same plough could be used for different activities, providing the necessary pattern of mouldboard was attached to it. Wheel and swing ploughs were discontinued when tractors came into large scale production. They required specially designed ploughs to meet the heavier forces.

The double furrow plough was first recorded in England by Walter Blith during the seventeenth century. Its huge wooden beam was divided into two parts, the first or fore section was long and curved up towards the horses, whilst the rear section was straight and sloped back to where its handles were attached. Each section was furnished with a share and coulter and cut separate furrows simultaneously, but it was exceedingly heavy and required a powerful team of animals before it. There was no great improvement in its form until the nineteenth century, by which time it had become a valuable asset to farmers who were tilling a greater number of acres in order to supply the increasing population. They naturally wished to economise on labour, and the double furrow plough, used wherever the soil was suitable, completed the work in about half the time taken by one single furrow plough; also two single furrow ploughs required four horses, but one double furrow plough required only three for exactly the same amount of work.

The nineteenth-century double furrow plough was made wholly in iron, either with or without wheels. Its beam was straight and larger than the single furrow plough in order to contain the second coulter, share and mouldboard. The double furrow plough was not suited to first ploughing on heavy land as it quickly tired the horses, and a single furrow plough went better. It was used with three horses for first ploughing on light soil and two horses for second ploughing, and could be adapted to subsoiling if the leading mouldboard was replaced by a subsoil tine. This plough, with or without wheels, readily lent itself to tractor haulage in the twentieth century.

Multiple ploughs could not be used to their best effect until the tractor was invented. A three-furrow plough was in existence by 1837 but it was

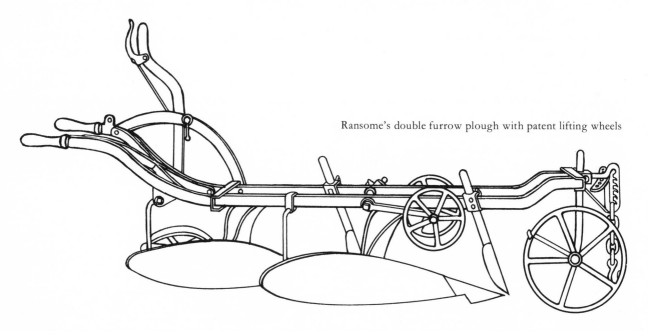

Ransome's double furrow plough with patent lifting wheels

not suitable for hard ground and so was adopted for work on light soil and for stubble paring in the autumn, when it proved a great deal more economical than a single furrow plough. Multiple ploughs, some turning as many as eight or twelve furrows at one time, were built for attachment to steam tractors and were mostly employed for breaking up the vast wheatlands in North America.

Turn-wrest ploughs were constructed in such a manner as to turn the furrow slices in one direction only. For this purpose two mouldboards were employed, each one being brought into use alternately, so that when the plough was moving towards the north, one mouldboard turned the furrow to the right of the plough and the other to the left on its return journey south, the effect being to leave all the furrow slices turned towards the east. These ploughs were particularly useful for hillside work and could quickly be turned around on headlands with less trampling than accompanied most other ploughs.

The Kentish turn-wrest plough was thought, by some people, to be similar to the Anglo-Saxon plough. The illustrated specimen differs little from those used in the eighteenth century. These ploughs were of powerful construction and consisted of a large pair of wheels similar in size to those of a waggon, iron rods linking the draught chains to the wheel axle and rear end of the beam, a curved beam some 8 to 10 ft. in length below which was dragged a thick slab of wood furnished at the fore with an iron share. These ploughs were pulled by oxen or a large team of horses, the number of which varied with the nature of the land to be ploughed. They were last used, but not on any wide scale, in parts of southern England during the early years of the present century.

By the end of the nineteenth century there were various styles of turn-wrest plough, mostly named after and reflecting the ingenuity of their inventors. Lowcocks of Marldon's turn-wrest plough contained two mouldboards placed end to end on the same side of the beam and was drawn from either end alternately. It was manufactured in number by Ransome and Co. The ploughman did not have to turn this implement at the finish of every furrow but simply unhooked the horses and attached them to the opposite end. This principle was probably not new, as it seems to have been employed in Devon towards the end of the preceding century. On Smith's turn-wrest plough the mouldboards were attached to a longitudinal bar which could be rotated in order to raise the idle mouldboard above the beam whilst the other was working directly below it. This idea was further exploited in the twentieth century and resulted in implements which required little effort on behalf of the ploughman to turn over or under, whichever movement was used to bring each mouldboard into use. Turn-wrest ploughs were essentially for use with horses and disappeared with the advent of oil tractors.

The balance plough, drawn by horses, was derived from the balance plough used with **cable cultivation,** but whereas the latter was a multiple plough and had a driver seated upon it, the horse balance plough usually turned one furrow only, with the ploughman walking at the rear. It took the form as illustrated with a main V-shaped beam balanced at the centre point upon a carriage and a pair of small wheels. Horses were attached by draught chains to the wheel carriage, and upon this point the beam could be raised or lowered in order to determine the depth of ploughing. Each end of the beam was about 4 ft. in length and provided

The Kentish turn-wrest or one-way plough was made very heavy in order to work the soils of Kent and East Anglia

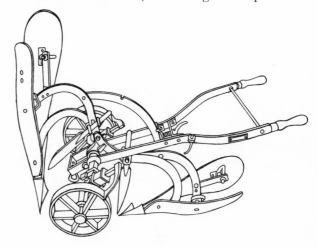

The horse-drawn balance plough, with two plough bodies placed heel to heel on the same beam

with coulter, share and mouldboard so that it could be pulled down into the soil whilst the other end was raised up. Each end of the beam terminated in a single handle which the ploughman used to steady the implement; a second handle was usually provided in the form of a steerage lever which extended from the wheel carriage. The lever was appropriate to both ends of the plough and was tipped over to the end in use. This plough was favoured by smallholders because it left little headland around the fields and, as the mouldboards were brought into use alternately, it laid all furrows the same way. Its use was confined to those areas where the land was flat, and it represents one of the last implements to be made especially for horse traction.

The paring plough generally replaced the hand **breast plough** in England and Scotland towards the end of the seventeenth century, although the latter was retained in less improved districts. The paring plough employed a flat share to remove a wide but shallow slice of turf along with other surface growth from land that was later to be ploughed up in preparation for seed sowing. The paring plough was brought to England from the Netherlands and for a long time was known as the Dutch plough or Dutch paring plough. In its original form it contained only one handle at the rear of the beam. This handle was peculiar in form, as it was provided with two short side handles or crutches projecting out horizontally on either side, the whole of which was probably carved from a tree growing to the required shape.

One crutch was near the end of the handle and served the user's left hand, whilst the second was lower down and served the other hand to turn and steady the plough at headlands.

The object of having a wide flat share was to cut a constant slice of turf which was then turned upon its edge by means of its mouldboard, where it was left for some time to the effects of the sun and wind. When dry it was gathered together into heaps and burned with the ashes being returned broadcast over the soil. Great care was taken not to overturn the turf; instead it was baked so as to crumble into small pieces convenient for spreading.

The practice of burning in order to improve land dates back to ancient times and most likely originated or was much used in the West Country of England, as it became commonly known as Devonshiring, Devonfiring, or Denbighshiring. The operation was not often repeated upon the same land, especially if it were shallow or stony, nor was it carried out on rich land, as it was believed to spoil the fertile juices. It was most frequently practised on land which had lain idle for a long time and had become sour or covered with undesirable growth, in the form of heather, gorse, or rushes. Burning in such situations increased the yield of the land by two or three times its normal amount.

By the beginning of the nineteenth century, the paring plough used in England was provided with two handles in the common fashion. Its wooden beam was about 8 ft. in length and raised up 1 ft.

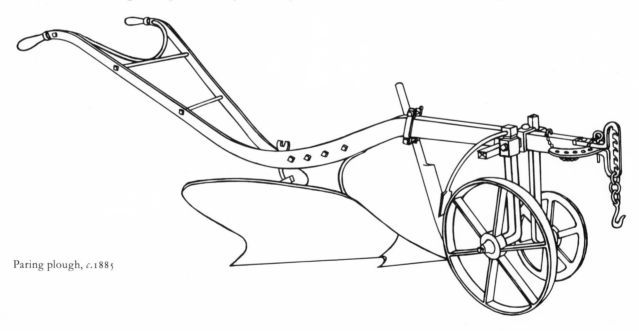

Paring plough, c.1885

above the ground, being supported at the fore-end by a disc-coulter and midway along its length by a flat, triangular shaped share, made of iron. The sharp edge of the disc coulter first cut through the turf in a vertical manner and prepared the way for the on-coming share, which was set to slice immediately below the turf and so separate it from the soil.

During the second half of the century, a large variety of paring ploughs were made in factories, but they were mostly laid aside before the end of the century as the necessity for burning the land was replaced by improved forms of fertilizer. They were however retained for the purpose of paring old pasture land in preparation for cultivation and for paring stubble land during the autumn, and can be obtained for attachment to a tractor.

straight lines was acknowledged and so was the after-use of a horse-hoe between the lines of growing plants. In the absence of any corn drill, grain could be sown in straight lines with the aid of a ribbing plough. The land had first to be prepared with harrows before parallel furrows were made, and using one horse on land in good order about three acres could be ribbed in one day, at intervals of 12 in. apart for wheat or 9 in. for oats and barley. The grain was broadcast into the furrows and covered by harrows, just as the ancient Egyptians did.

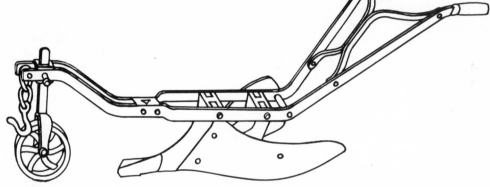

The horse-drawn ridging plough was used when the crop was showing above ground. Employed between the rows, it pushed the soil into ridges around potatoes, celery, etc.

The ridge plough was built to different patterns before the nineteenth century. An early form consisted of a single spoked wheel about 24 in. in diameter, with two wooden handles about 6 ft. in length extending back from its axle. The operator used the handles to push the wheel before him and by means of a pointed, iron blade, situated in front of the wheel, he penetrated the soil and raised it up equally to either side in the form of ridges, leaving a hollow into which seed was sown broadcast. The majority of seeds fell into the furrow and were covered by subsequent harrowing. At the same time there existed a simple wooden plough for the purpose of forming ridges. It was similar to the later, nineteenth-century version except that it lacked the high double mouldboards which the latter had attached on either side. Such ploughs were first made in the Lothians for some years before they were introduced into the East Riding of Yorkshire about 1800. They were called ribbing ploughs and formed a substitute for corn drills which were expensive and outside the means of many small farmers. The benefits of sowing corn in

Horse-drawn ridging ploughs, with and without wheels, were widely used for earthing up potatoes and other growing root crops, and so came to be known as moulding ploughs in many districts. The northern counties of England, especially Lancashire, preferred a full-breasted or less concave shape of mouldboard than was usual and, with the addition of two side wings attached to the share, it formed higher ridges with steeper sides than the common pattern of ridge plough. They were often used for opening water furrows and, with both mouldboards removed and a single broad share in their place, were effectively used for stubble skimming in the autumn. Three-row ridgers became popular in potato growing districts, early this century.

The gallows plough, drawn by horses, was popular on the Continent, but in England it was used mostly in the West Country, during the nineteenth century, for deep work and for ploughing in coarse manure which would quickly choke other kinds of plough. For this reason the beam was raised exceptionally high at the fore-end where it rested upon the cross-bar or saddle of a wheeled fore-carriage. This consisted of two wheels usually about 3 ft. 6 in. in diameter which were set at either end of an axle. The fore-carriage was free to turn, as it was attached to the beam by two chains only, which came back from the wheel axle and were secured to the beam near where it joined the stilts. Hardly any direct steerage was required from the ploughman walking behind, since the rear end of the beam and the plough share were forced, by the two connecting chains, to follow the movements of the fore-carriage which was attached directly to the horse. The light Norfolk wheel plough had a gallows, but the type was discontinued with the passing of horse traction.

The sulky plough. The term 'sulky' was used by American farmers when they referred to a single furrow-wheeled plough upon which the driver was seated. This type of plough was commonly called a riding or pole plough in England, where it did not become at all popular whilst horses were the source of power. It was believed to have greater draught than other ploughs, but nevertheless two British manufacturers sold them in small quantities after 1881. Patents were taken out in America from 1844, for early forms of sulky ploughs with two wheels, and forty years had to pass before a third wheel was added. One of the first two-wheeled sulkies to be made practical and manufactured on any scale was the one invented by F. S. Davenport, for which he received an American patent in 1864. Even if this plough did not sell in such great numbers as the one derived from it by fellow countryman Robert Newton of Jerseyville, Illinois, it certainly captured the imagination of the trade. Newton's improved plough included a disc coulter, which thereafter became a standard feature, and the use of a single plough body instead of two. Several American manufacturers had made their versions available by 1870, and at the public trials held at St Louis in 1873 there were no less than sixteen different makes of sulky ploughs competing. At this time both the furrow wheel and the land wheel were set vertically on either side of the plough and the weight of the driver and framework were supported by the sole of the plough, which was made to slide on the unploughed land and was extra long for this purpose. This was improved upon in 1876 by W. L. Casaday, who was granted a patent for a two-wheeled sulky which was eventually made in large numbers by the Oliver Chilled Plough Works at South Bend, Indiana. This inventor was the first to place the furrow wheel at an angle so that it inclined against

'Flying Dutchman', a three-wheel plough introduced by the Moline Plough Co. From R. Ardrey's *American Agricultural Implements*

the side of the furrow and thereby ensured that the sulky remained stable whilst cutting a furrow of even width. This idea of the inclined furrow wheel allowed the long sole to be dispensed with and was continued as a feature in three-wheel sulky ploughs developed in the following years.

The flying dutchman, pioneer of three-wheel sulky ploughs, was introduced by the Moline Plough Company, U.S.A. in 1884. It was made with two furrow wheels, set in tandem, one at the fore-end of the beam and the other at the rear end, with a larger wheel extending out on the centre land side. The driver was seated directly above the rear furrow wheel, and from this position he could observe the 'slice' being turned over whilst he could also provide any adjustment to the working depth or lift the plough out of work, whenever necessary, by a compound lever device attached to the beam. **The little yankee** was a famous three-wheeled sulky plough developed by the Grand Detour Plough Company, who had previously earned a reputation with their self-scouring prairie ploughs and two-wheeled sulkies. It was the first three-wheeled plough to be really stable when working. This was due to both furrow wheels being inclined against the land, the rear one being more or less inclined by means of a lever attachment. Instead of being hitched to horses by the framework, as were former ploughs, it was drawn from the end of the beam. This improved manner of hitching allowed the fore-end of the

'Little Yankee' developed by the Grand Detour Plough Co. From R. Ardrey's *American Agricultural Implements*

They were the forerunners of today's tractor ploughs. **The riding plough** was drawn at the rear of a tractor on a length of chain and, being separate, it required the attention of a driver, who was provided with a seat at one end of the beam, where he was able to observe the work in progress. On general farmwork, the plough turned two or three furrows and was of convenient size for the driver to reach out from his position and adjust the furrow, land and hind wheels through different levers. His extra weight and constant supervision were considered to be an advantage especially when working on rough ground. Riding ploughs, for turning a large number of furrows, were built in England at the beginning of the twentieth century; the majority of them were sent to America and Asia, where they were hauled by powerful steam tractors and used for breaking up new land. They were huge in comparison to those used by the average farmer and normally had ten or twelve plough bodies attached to the beam, each one being provided with a separate lever so that the driver could regulate its working depth. He was provided with a narrow platform which ran the whole length of the plough and allowed him to attend the row of levers whilst the ploughing was in progress. The plough bodies, which contained the mouldboards and shares, were set well below the beam and were placed wide apart in order to avoid continual blocking, and disc-coulters were normally provided on such ploughs for the same reason. This type of plough was gradually replaced by the self-lift type.

The self-lift plough was in general use by the mid-1940s. It was attached to the tractor by a short draw-bar and could be controlled by the tractor driver from his seat. The various levers or screws to regulate the work were placed within easy reach of his hand, and it was also possible for him to reach back and throw a lever which caused the plough to climb gradually out of work, where it was held clear of the soil until released. If the driver's seat was not in a convenient position, then the lever was replaced by a cord which he pulled in order to activate the lifting mechanism. Various lifting arrangements were devised by manufacturers. The most common employed a toothed rack made to engage with a pinion on the nave of one of the land wheels. The wheel was contained on a cranked axle which, through the engagement of rack and pinion, was moved from near horizontal to vertical position and lifted the plough out of the ground.

plough to be raised or lowered on the furrow wheel standard and afforded the driver a much greater measure of control over his implement. It also bettered the adjustment for depth and width of furrow, and the hitching could be altered to adapt the plough to hard or soft ground.

Ploughs are also used for drainage work. **Subsoil ploughs** for breaking up the close hard soil below the furrow, and **mole ploughs** for land drainage work are described in the section on Land Drainage (see page 17).

Early tractor-drawn ploughs varied enormously in shape and size but could be divided into two basic types, riding ploughs and self-lift ploughs. They were expensive items, so rather than have a number of separate ploughs for different purposes, the majority of farmers purchased a general purpose type or obtained such a model as would allow interchanging of various parts. They were built a good deal longer and heavier than any previous plough and were usually fitted with wheels, a landwheel which travelled on unploughed land and regulated the depth of soil penetration, and two furrow wheels, one at the fore-end of the beam and a hinder furrow wheel if it was a three or more furrow plough. With the increased power of the newly developed tractor, manufacturers designed these ploughs to turn a large number of furrows simultaneously whilst working at an increased speed.

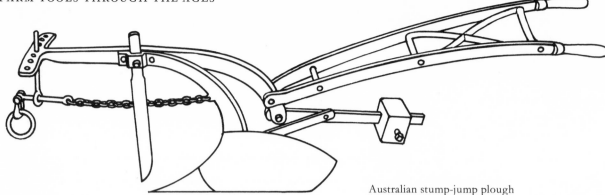

Australian stump-jump plough

The stump-jump plough was invented by Robert Bowyer Smith of Adelaide, Australia, in 1876 and accommodated the presence of tree stumps and large stones buried in the earth, which defeated the work of imported British ploughs. His first stump jumping plough was the three furrow 'Vixen', designed to jump or ride up over obstructions and then resume its correct position. Smith's idea was scorned by many Australians, but in the same year he proceeded to build an improved single furrow plough on the same principle as his Vixen. It closely resembled a common swing plough in appearance, but its mouldboard and malleable iron share were not rigidly fixed below the beam in usual fashion. Instead they were suspended in an iron bracket which was hinged to the beam by a single bolt, so that whenever a buried obstacle was encountered, the share would give way and cause the plough to jump up or ride over it. The share was forced back into its working position $\frac{1}{2}$ in. behind and above the point of the coulter by the use of a weighted lever arm; this too was attached to the bracket and exerted a pressure of up to 500 lb. against the point of the share. This plough was exhibited and seemed to avoid the abrupt collisions which resulted in broken shares and damage to others. It was eventually manufactured on a large scale and the principle since applied to many other implements of Australian origin. Stump-jump ploughs have also been used in America, particularly when breaking up new land.

The disc plough employs a steel disc which does not completely invert the furrow or bury surface weeds as efficiently as the ordinary mouldboard. The disc has varied between 20 in. and 30 in. in diameter and is inclined at an angle against the furrow. The disc plough has been of great value in hot countries where the stubble land is too hard for other types, and it performed equally well in wet sticky land where other ploughs would not scour. Disc ploughs

were produced in England as early as 1870, by Messrs Corbett of Shrewsbury, but they were not widely used except for stubble skimming. It was in Australia, America and Canada that most progress was made in the design of the particular implement, and during the early years of the twentieth century it developed from a single furrow riding plough to a multi-furrow plough for tractor attachment. John Shearer and Sons, Australian plough manufacturers since 1877, were one of the first mass producers of the Stump Jump Disc plough. Their 'Sovereign' plough ranged from ten discs to twenty-four, each 23 in. in diameter and $\frac{3}{16}$ in. thick. Set at the end of the spring loaded arms, each disc would jump a maximum height of 12 in. Disc ploughs are still extensively used in hot countries where the soil would defeat other kinds of plough share.

The American plough. Progress in the development of American ploughs was slow before the nineteenth century and both the design and use of the wooden 'bull plough' was haphazard. The common practice was for the farmer to make his own

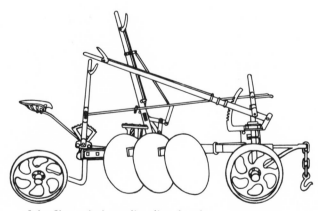

John Shearer's Australian disc plough

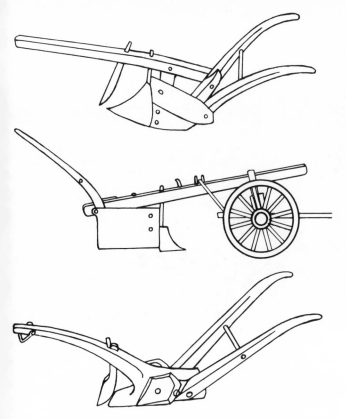

Top and centre: Ploughs used by American pioneers
Bottom: Australian pioneer plough

1819, but although it offered a scientifically designed mouldboard and standard interchangeable parts, it was rejected by farmers, as they were of the notion that 'cast-iron poisoned the land and stimulated the growth of weeds'. The old 'bull plough' with its wooden mouldboard was still as popular as ever, and it was only after a considerable waste of time and expense that the objection to cast-iron was overcome. Then farmers in New England and along the eastern shores acclaimed Wood's cast-iron plough, which was adapted to various kinds of soil. However the cast-iron plough would not scour properly in the prairie soils of Illinois and a considerable amount of pioneering work remained to be completed before a suitable plough was developed. The early settlers had already achieved a great deal, having cleared away timber and opened up the prairie land to reveal the rich virgin soil beneath. Their original implements had proved inadequate to the task and so better ones were contrived, amongst which there was the prairie breaking plough of great strength and drawn by two to six oxen. It was designed to rip open thick prairie grass and matted turf and was built on similar lines to the common wooden ploughs, but was larger and heavier; the mouldboard was higher and more sweeping, whilst the share and coulter were edged with steel to prevent undue wear. The wooden mouldboard was replaced in about the middle of the nineteenth century by a skeleton mouldboard composed of curved iron rods, which Australian farmers also found an asset when cultivating Wimmera Blacksoil.

Pioneering work was not finished in Illinois even after the prairie breaking plough was perfected. Wooden or cast-iron ploughs could only be used for cultivating with great difficulty, as they did not scour after the light vegetable mould had been worked for two or three seasons. After much trial it was discovered that a plough with a mouldboard made in steel would scour quite readily and perform better than cast-iron or wood. The first steel plough to be recorded in America was made by John Lane in Chicago, during the year 1833, at a time when suitable steel was unobtainable, except in the form of carpenters' saws. He covered the mouldboard of his first plough with the blade of a long steel saw, which he first deprived of teeth and divided into four portions, three of which covered the face of the mouldboard, whilst the fourth was used for the share and the triangular shin piece behind it.

and obtain the necessary iron parts from a black-smith, or alternatively he purchased the complete plough from a local plough maker who probably performed as a wheelwright and carpenter as well. The ash wood beam and stilts were straight and brought together with the other parts in accordance with neighbourhood convention rather than with regard to the nature of work. The resulting plough consequently required the user to exercise special skill to use it effectively. The mouldboard was likewise carved in wood, but in order to prevent it from wearing away too quickly, pieces of sheet iron, horseshoes or hoe blades had to be nailed on to the face. In 1788, the renowned statesman Thomas Jefferson followed the work of James Small and published recommendations for a scientifically designed mouldboard. Some improved ploughs were made for him and distributed amongst farmers as a means of encouraging a general acceptance, but they were spurned and much of Jefferson's valuable work was wasted for almost a generation. Jethro Wood patented his cast-iron plough in 1814 and followed it with an improved version in September

Deere and Co.'s three-wheel plough. From R. Ardrey's
American Agricultural Implements

The situation remained unchanged for many years and the only steel available was in the shape of saws, all of which were procured, along with much coveted blanks, from a manufacturer, and were used for plough-making. Eventually the required steel, rolled to a width of 12 in., was obtained from Pittsburg. Then John Lane and others commenced their business in earnest. The demand for steel ploughs was overwhelming and is illustrated by the production of John Deere's ploughs first patented in 1837. Forty were sold in 1840, 400 in 1843, 1,000 in 1846, and by 1857 his extensive factory at Moline, Illinois, was producing some 10,000 annually. More makers entered the field, amongst whom was Joel Norse, who in 1842 moved from Shrewsbury, Massachusetts, to Worcester, where he presented the first in his series of Eagle ploughs. It was considered to turn a better furrow than Lane's plough, perhaps due to its longer mouldboard. In 1866 John Lane achieved further success with the perfection of a

'soft-centre' steel for shares and mouldboards. Other inventors had been at work on the same idea; in 1862 William Morrison had produced a soft iron faced with steel plate, but his material was inclined to warp when the parts were tempered. But Lane composed his plate in three layers, the outer ones being steel with a centre of soft iron, and distortion was avoided as one layer of steel counteracted the other when the parts were heated. They retained the desired curvature and scouring qualities whilst the iron centre afforded considerable strength.

The following testimony to American plough makers appeared in the *British Farmers Magazine,* of 1859:

In the Black vegetable loam of the Western Prairies, the cast-iron ploughs used in the heavier soils of the older States will not cleanse, so in their stead the farmers use light steel breasts that, when turned out from the factory are polished bright as silver. At the Western shows I saw collections of 200 perhaps of these beautiful ploughs: the wood of the beams and handles is selected with care, sand-papered and varnished and I assure you it was a sight to look at. There was scarce a plough in the lot that would not make an ornament for a shop window and then at the trials, to see twenty or more of them turning over the black soil in the most efficient manner would gladden the heart of Allen Ransome.

The dynamometer was a simple instrument used for measuring the draught of ploughs. It was invented by Mr Samuel Moore, Secretary to the Royal Society of Arts, and was first employed by the Society at their competitive trials during the 1780s, where it brought to light the enormous difference that existed in the draught of ploughs. The Society was prompt to encourage all farmers to obtain a dynamometer, by which they could ascertain the correct plough for their type of land, and many of them were surprised to find that the plough they were accustomed to using could be bettered. The common dynamometer was a spring-operated instrument, small enough to be carried in one hand and inserted by hooks at either end into the draught chain between horse and plough. As the horse pulled forward the whole of the tension was thus passed through the meter, its steel spring distorted and the draught indicated by the pointer upon its

Simple dynamometers

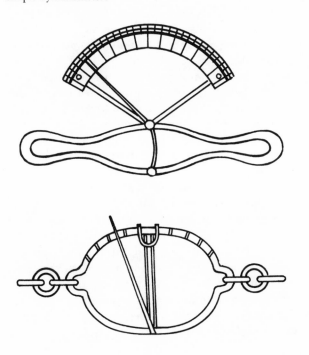

dial. The general arrangement of two such dynamometers is illustrated. Both of them employed a near circular spring about 6 in. in diameter and whilst the majority of other dynamometers worked upon this principle and indicated with sufficient precision for the average user, some difficulty was experienced when trying to take a really accurate reading by eye, due to rapid fluctuations of the pointer when the plough passed over rough ground or the horses faltered. The problem was rectified by the French invention of a self-recording dynamometer during the 1850s. It differed from the former type in that intricate mechanism was contained inside a narrow box about 24 in. long raised up on a wheeled carriage into the line of draught. The variance of traction between horse and plough was transmitted through a coiled spring which fluctuated and caused a graphite marker to scribe a continuous undulating line upon a moving web of paper, thereby recording a permanent and accurate description of the power required in drawing the implement from start to finish. Dynamometers have lately been replaced by sensitive electronic measuring apparatus.

The plough sledge. Difficulties encountered on removing a horse-drawn plough from one part of the farm to another were overcome in the early years by the use of a simple sledge which moved on iron shod runners. Towards the middle of the nineteenth century it had become a low triangular bogey with two small wheels and a handle for pulling. The plough could only be raised a few inches above the ground, but that was sufficient to avoid the damage which would occur if it were dragged along the road. Tractor ploughs did not require such a sledge.

The plough staff. Before tractor-drawn ploughs were developed, it was essential for a ploughman to carry with him a staff, in order to clear his implement of any roots or weeds that might accumulate before its coulter or around the wheels, and also occasionally to scrape the mouldboard. The staff was fixed by hooks to one stilt of the plough when not in use or, if the conditions were unfavourable, the ploughman held it ready in one of his hands that steered the plough. The plough staff had a blade that resembled a miniature spade, joined by a socket to a wooden helve some 5 or 6 ft. in length. In the latter years of the nineteenth century the helve of the plough staff was made in a lightweight tubular metal, the convenience of which enabled the ploughman to wield and scrape with far less effort.

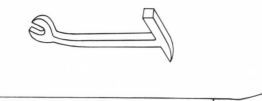

The ploughman's tools: spanner and staff

The plough spanner. It became customary for manufacturers to supply a spanner with each of their ploughs. Shaped in iron to a size convenient for one hand, it comprised at the same time both spanner and hammer. It had a cross-head with a claw at one end of the handle for hammering and a socket for turning nuts at the other. A staple was provided on the beam or on one stilt of the plough for its location, the spanner being prevented from falling through by the cross-head. In such a position it was readily available at any time for adjustment to the share or coulter.

A marking staff was used to straighten and govern the distance between ridges. It was an iron rod about 2 ft. 6 in. long pointed at the lower end and secured in a vertical position by a bracket on the side of a plough beam. It scratched a line parallel to the furrow made by the plough and as this line was plainly visible it served as a guide for the straightness and width of subsequent ridges.

The brander. By the middle of the nineteenth century there were a number of devices available which could be attached to a common plough for the purpose of taking up potato crops. The brander was a name given to one such plough attachment, reputed to turn out potatoes without the slightest damage to the crop. It was patented by Messrs Lawson of Edinburgh and consisted of iron rods about 3 ft. in length joined together. In order to raise the potatoes, the mouldboard was removed from a plough and the brander bolted onto the framework in its place. It then lay immediately behind the share with its widest end towards the rear, and its effect when the plough was moving forward was not to turn the soil aside as any ordinary mouldboard would have done, but to allow it to pass between the rods, whilst the potatoes being larger than the particles of soil were stopped by the rods, from where they fell down onto the newly turned soil. The haulms had first to be cut away from the plants and removed from the field before work commenced in order to prevent the implement becoming choked.

The plough graip worked on a similar principle to Lawson's brander but was intended for use with a **double mouldboard plough** and its particular action of turning the soil to both the left and the right hand sides simultaneously. The graip was formed by four iron bars which radiated from an attachment stem, the two outside bars being 28 in. long and $1\frac{1}{2}$ in. wide, the two inner bars $19\frac{1}{2}$ in. long and 1 in. in diameter. Its working position was in the centre of the plough, behind the double mouldboard, where the narrowest end of the graip was attached by two bolts. In order to harvest potatoes, the plough was drawn along the length of the potato ridges with its share penetrating well beneath the roots. A mixture of loosened soil and potatoes was thrown up and then directed by the two mouldboards to the rear of the plough, where both immediately fell upon the radial bars of the graip. The soil passed through whilst the potatoes were retained for a brief moment before they rolled back along the bars and fell clear of the moving plough. Any small potatoes that had passed through the graip, and those not initially turned out of the soil by the plough share, would have remained buried had not the last few inches of each bar been pointed and turned down into the soil. These points acted as a harrow and, by disturbing the soil, they brought the hidden potatoes to the surface.

The earliest form of mechanical **potato digger** appeared in 1852 and a practical rotary potato digger followed three years later. They were designed to speed the harvest and had generally eclipsed the potato raising plough and plough attachment by the end of the nineteenth century.

The ristle. An implement used in some parts of Scotland before the nineteenth century was known by this name. It was usually employed in conjunction with the old Scottish plough until that type of plough was laid aside: but it continued in the less progressive Hebridean Islands for many years. Its function was to slice through turf before an oncoming plough and also to clear the way for digging. It was not unlike the primitive ploughs in appearance, having two handles extending from a large wooden beam below which was fixed a sharp iron blade. Two men were required to hold the handles at the rear and horses were harnessed to the front so that when the ristle was working, the weight of the beam lay completely upon the blade.

The subsoil pulveriser, a departure from the **subsoil plough** was invented in 1841 by a Mr Nugent of Hinckley, Leicestershire. Its function was similar to the subsoil plough in that it undermined and stirred the subsoil without mixing it into the layer above, but whereas the subsoil plough was particularly suited to deep draining work, the pulveriser was better used on land that had become consolidated into a hard packed mass, by the trampling of horses, through which roots and rain water could not penetrate. The underside of its beam was furnished with four curved iron tines to operate in succession between 6 and 16 in. in depth, stirring and pulverising to that extent and to a breadth of 12 in. In this manner the sub-strata was not disturbed by a single share, as with the sub-soil plough, but was shattered by the successive action of the several tines, each working at a different depth suitable to the soil and conditions. The pulveriser was equipped with two small wheels at the fore-end and two larger ones at the rear, with a lever adjustment on the axles to raise and lower the beam, the consequence of which was to alter the working depth of the tines.

A turn-wrest snow plough was one implement that resulted from the ingenuity and experiment of the nineteenth-century inventors. It was of great assistance to farmers in the higher regions of England, Scotland and Wales, being used there to clear away snow along the hills and mountainside, leaving tracks for the convenience and preservation of their sheep. This mountain snow plough was a huge triangle of iron, lying horizontal in the snow, with two sides 7 ft. 6 in. long, $1\frac{1}{2}$ in. thick and 15 in. in depth, whilst the third side was a single iron bar, 6 ft. in length, attached to the ends of the longer sides. This bar formed the hindmost part of the plough and to it were attached one or two movable stilts, which the ploughman used for steerage or pressed down upon when necessary to prevent the implement from rising up. The apex of the triangle was immediately behind the horses, who were joined by a draught chain to this point. Its effect when drawn through deep snow, would have been to cast the snow on both sides, but in order to avoid this and cast the snow to one side only, that being down the mountain slope, a movable head was attached to the point of the triangle. This was also of iron, $1\frac{1}{2}$ in. in thickness and 15 in. in depth, made into an angle with one side 18 in. and the other 30 in. in length. It was attached by two bolts at either end to the front of the plough, serving to break the symmetry. When working across the slope of a mountainside, with the right hand side of the plough nearest to the bottom of the slope, the head was attached with its longer side to the right and thus threw the snow in that direction. At the end of its run, before the plough was reversed for its return journey, the head was taken off, turned over and refitted with its longer side to the left of the plough, which would, on returning, be nearest to the bottom of the slope and so continued to move the snow in that same direction.

Implements and Tools for Cultivating the Soil and Maintaining the Land

The caschrom is one example of a hand plough and is thought to have been devised in the Middle Ages or even later, but this is by no means certain. However, despite improved methods of ploughing, the caschrom was much used in the Hebrides until the early part of this present century. A curved handle some 5 ft. long was selected from a tree growing to a naturally convenient shape and after being trimmed was fitted at the base with a foot peg and a long sharp wedge which pointed forward. The worker would hold the caschrom vertical with both hands and rest the end of the handle upon one shoulder. By kicking down into the foot peg he would force the point of the wedge into the soil. Pulling the handle backwards he would tear up the soil, throwing it to either left or right, and proceed to work backwards until the whole area was turned over. This instrument is probably still used by some Hebridean crofters, but only for cultivating small strips of land. An expert user would complete twice as much work in one day as a man working with a spade.

The breast plough. Until late in the eighteenth century this implement was necessary for paring off turf and other growth before land ploughing commenced. The iron blade was of a similar shape and dimension to the **turf spade**, being angular and sharp-edged with its right side turned up about 3 in. The blade was attached at an angle to a stout wooden helve about 6 ft. in length. This helve was flattened towards the end nearest the ploughman and was tenoned into a cross-handle which rested against his chest or thighs as he pushed the spade along with the blade below the turf. He pared the turf in slices about $\frac{1}{2}$ in. thick, but the task became excessively hard labour where the land was covered with rough top growth and the soil full of stringy roots. Then it required a strong man to pare the turf thicker and turn it over as it was cut. Paring was usually carried out by a team of labourers who followed each other, as when hoeing or mowing with scythes. They raised the turf in one direction only and laid it on edge, so that it was exposed to the drying effects of sun and wind. After two or three weeks had elapsed it was burned in small heaps and the ashes ploughed into the soil. In Scotland this implement was called the flauchter spade and considered to cut turf thinner and more uniformly than the **paring plough,** by which it was generally replaced in England in the early nineteenth century.

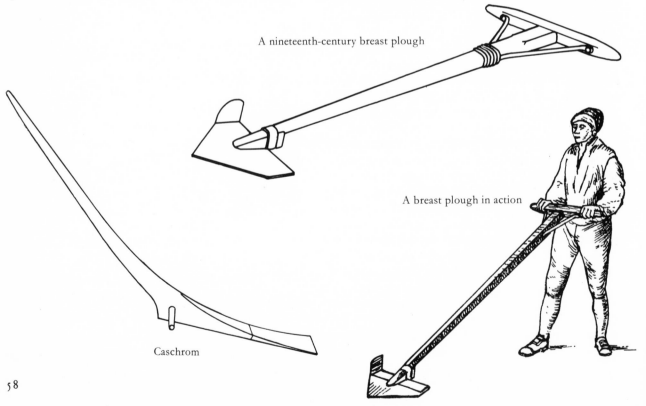

A nineteenth-century breast plough

A breast plough in action

Caschrom

The paring mattock. In the west of England, the labour of breast ploughing was aided by a paring mattock, in some localities called a beating axe or cobbing hoe. It was finished in the manner of a hoe, with a single iron blade attached to the end of a stout helve at an angle of 45 degrees. The blade was some 12 to 14 in. in length, 6 in. across its cutting edge, and tapered upwards to the helve. The labourer used this instrument in an operation called spading the ground, during which he hacked and cleared the roots of gorse and other obstinate growths which were likely to hinder the breast plough.

The double bitted mattock was employed in like manner and was identical to the one described except that its blade was extended on the opposite side of the helve for 2 or 3 in., where it was fashioned into the form of an axe. These hand instruments were retained after the breast plough had passed, as they proved to be useful with the horse-drawn **paring plough.** There is seldom any call for this type of tool in modern farming, though various mattocks are still obtainable. The early tribesmen, Egyptians, Japanese, Britons etc. all had mattocks of some kind.

The digging mattock was a type of pick used by farmers and others to loosen stones, stiff clay or hard chalky land, prior to it being turned over or removed with a spade. Handled skilfully, it could raise enough earth in one single movement to half-fill a horse-drawn cart. Its shape varied according to district and maker but the common pattern, used by agriculturists in the seventeenth century, was fashioned with an iron blade about 16 in. in length, set at right angles to a strong wooden handle. The blade was flattened out to a sharp adze-like edge about 5 in. wide in order to penetrate the soil and cut through roots. When loosening a quantity of earth, the user first had to dig out a trench about 12 in. in depth, then moving about 3 ft. away, he would, with the aid of a **beetle,** drive the blade straight down into the ground until it was prevented from going deeper by the handle. Two men were then required to raise up the end of the handle which caused the blade to lever up earth in the form of large lumps for as far as the trench first excavated. The operation was repeated in similar manner at intervals alongside the open trench until enough earth had been loosened. This mattock remained useful until it was replaced by more efficient digging instruments in the nineteenth century.

The clodding mell was used in England before the sixteenth century to smash down large clods of earth left by the plough. It was a heavy wooden mallet, the hammer-end being 12 in. or thereabouts in diameter, probably made of apple wood and bound round with straps of iron. Sheer physical strength was the only quality required of its user, who during the eighteenth century was gradually relieved of the exhausting task by the introduction of horse-drawn rollers. They were more effective as clod crushers and had completely replaced the clodding mell by about 1800. However a large wooden mallet, commonly known as a 'beetle' is still extensively used for driving fence and hurdle posts into the ground.

A thirteenth-century engraving depicts an English farmer holding the handle of his plough with one hand and wielding an axe-like instrument in the other. It appears to have a sharp edge for clearing away roots and a blunt side for smashing down clods. Improved harrows completed the task of breaking soil after about the seventeenth century.

The clod knocker was improvised by a Rutland blacksmith who flattened and joined the prongs of a pitchfork. Such an instrument was used after the passage of a horse-plough for the purpose of lifting, shaking or smashing clods of soil.

Hand hoes were employed for the destruction of surface weeds growing between rows of corn or

Nineteenth-century hand hoes

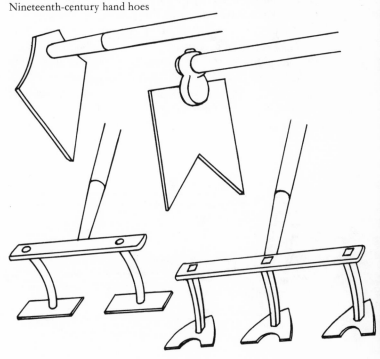

green crops, for loosening the soil and for 'thinning out' plants sown in rows or by hand broadcast. The illustrated hoe-heads indicate the wide variety of patterns employed over the years. The earliest hoes were forged from one unbroken piece of iron, whilst later ones had detatchable blades which could be taken out for sharpening or replacement. The head was set at the end of a wooden handle (or stail), the length of which varied between about 3 and 5 ft. There were two distinct types of hand hoe, the draw hoe which the user pulled towards himself and the opposite type of Dutch hoe which he pushed away. Various designs of hand hoe are still available for the gardener, but they are seldom required by the farmer.

The draw hoe, with which freedom to make long strokes over a prolonged period was essential since it was commonly used in the field to loosen the surface of hard soil and to clear away weeds from root crops and corn that had been sown in rows. Different widths of hoe became available to suit the various crops, and often a swan or crane-neck shank was used to join the blade to the handle, the curved form of which acted as a spring between the two wherever pressure was applied. When hoeing a field of corn it was necessary for a number of labourers to arrange themselves in the form of a chevron so as not to obstruct each other when working backwards along the line of the drills. The hoe for thinning turnips was made with a short handle so that the user was forced to stoop low and observe his work closely. The draw hoe can readily be purchased to the present day.

The Dutch hoe is nowadays without the short cross-piece at the end of its handle which was a common feature of its early form, and the blade is frequently rectangular in shape. The early blade pattern was triangular, made in iron with sides about 6 in. long, and its extreme point was used to dig up the weeds by penetrating beneath their roots and also for breaking up soil between plants where the use of a spade was undesirable. The user grasped the hoe in both hands and jabbed at the soil before him.

Hand rakes were not used much for the preparation of land or the covering of seeds, as horse-drawn **harrows** would complete this work quite readily.

The drill rake occasionally used on small farms was of heavy construction with a wide head to facilitate the drawing of drills for small seed. The wooden handle was some 5 ft. long and the head, made of iron, was about 2 ft. wide. The short iron teeth were triangular in section and were set at regular, large or small, distances apart. After reducing the top soil to a tilth, the user drew his rake along, with the handle almost horizontal, to incise shallow drills for the seed. Hand rakes of various forms are still manufactured for the market gardener.

The daisy rake was peculiar to England and was used during the nineteenth century to rip the heads from daisies, buttercups, dandelions, and other plants that flower in short grassland. Its wooden handle was about 5 ft. long and joined the centre of a head some 2 ft. 9 in. wide. The tines of this rake were made of thin plate iron, each some 6 in. long and $\frac{1}{4}$ in. thick. The upper half of each tine was $1\frac{1}{2}$ in. wide tapering down to a point. About twenty of these tines were fixed by their top edge along the head of the rake with a slight gap between each. They were also made to curve under for the last few inches of their length. The farmer took this rake in both hands and proceeded to walk backwards or sideways pulling the rake across the surface of grassland. Grass and other growth caught up in the spaces between the tines would normally pull through without damage, but should the growth be in flower, then its head was too large to pass through the space and was ripped away from its stem. The introduction of the horse-drawn **mowing machine**

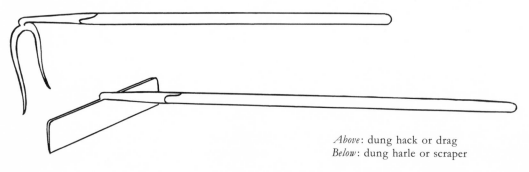

Above: dung hack or drag
Below: dung harle or scraper

in the nineteenth century meant that this rake and other implements for removing weeds from the field could be laid aside.

The dung hack, simplest form of rake used on the farm, consisted of two or three sharp prongs, each about 8 in. long, at right angles to the handle. It was some 8 ft. or more in length and made to assist the withdrawal of dung, in measured quantities, from a moving cart by a labourer walking at the rear.

The harle, so called in the north of England, was an iron scraper fixed at right angles to the end of a wooden handle. The blade, about 18 in. in width and 3 in. in depth was convenient for scraping up mud and other wet matter in the farmyard or barns. Both hack and harle are still to be found on many farms, where they serve as general purpose tools.

The digging spade, nowadays common to both farm and garden, is very well known, but there was once a multitude of shapes and patterns, all of which were forged locally. Fenland farmers found a pointed blade, sharpened like a knife around its edge, well suited to digging soft, marshy land, and for cutting the roots of water weeds. A similar sort of spade, but with one side edge turned up vertically for an inch or so, facilitated the paring of turf and other top growth. Iron digging blades were generally tapered down towards the cutting edge and the wooden helve was received into a socket on the top edge of the blade. There was often a short iron foot-tramp upon which the user set his foot when digging. The tramp projected out at right angles to the helve and was joined to it by a swivel ring so that it could be set on either side according

to whichever foot exerted the pressure. These spades, suited to hard lumpy clay, must have been equally efficient for digging deep land drainage trenches until they were replaced by a selection of narrow draining spades in the nineteenth century.

Spades were often called after their place of origin. The Portuguese spade had a pennon or swallowtail blade well suited to deep digging and penetrating between stones, where a common straight-edged blade was defeated. The Cornish spade is still used on some farms in the west of England and in Wales. It has a heart-shaped blade with a wooden helve about 4 ft. long set into it at a slight angle. The Flemish spade was long and scoop-like. They were all descended from primitive spades used by early tribesmen, and some bushmen still use wooden spades most effectively.

A seventeenth-century pattern of spade, with its sharp blade fashioned like a crescent moon, was used by horticulturists in the midland counties of England for cutting the edges of turf, and because it was particularly suited for that purpose farmers in the same area found it a convenient instrument for replacing any turf damaged by colonies of ants. Before their numbers were reduced by successive ploughing, ants were a considerable nuisance, and any farmer who allowed them to settle in his meadow or pasture lands would eventually find his crops

from left to right:
Open spade for digging clay
Portuguese spade
Iron plate laced beneath the worker's boot as an aid to digging clay or hard soil
The horizontal spade, for making footings etc.

61

wasted and his scythe obstructed by their hills. In order to destroy as many ants as possible the hills were levelled during winter. The soil was taken off in two or three layers and scattered about, after which the core was dug out and crushed beneath the farmer's boot. The hole was left open until the beginning of spring so that the frost and rain could penetrate, and when the turf was finally replaced it was made to lie lower than the surrounding ground in order to retain water and discourage the ants returning. The half-moon pattern of spade is still widely used for turf culture, but it is much lighter in weight than the seventeenth-century version.

The Flemish spade was probably brought into England from the Continent during the late eighteenth century. The Flemish helve was about 8 ft. in length and was designed to release the operator from continuous stooping whilst at work. The blade was slender and concave, in the manner of a scoop, with sharp edges running to a point in order to pierce the soil easily. The blade was not provided with a foot tread as the unusual length of the helve enabled enough pressure to be exerted from the shoulders alone for digging or moving soil. Loudon mentioned it in his 1836 *Encyclopedia*, but its use does not seem to have extended beyond that time in England.

The dung spade was used to cut compressed manure into conveniently sized pieces, when removing a heap from the farmyard to a field. The most prominent feature of the dung spade was its heart-shaped blade which was about 18 in. in length and 12 in. wide, fashioned to a point and kept sharp around its full perimeter. The helve was about 18 in. in length and was provided with a cross-head of the same dimension. To use the spade it was grasped by the cross-head and thrust down as far as possible into the heavy, compressed dung; then withdrawn and thrust in again as many times as were necessary to cut the dung into suitably sized blocks. These were lifted by a graip and thrown into a cart. A spade of this type was essential when animals were kept in quantity.

The hedge spade was designed particularly to assist the worker in clearing hedge-banks or other steeply sloped land of grass. The blade was small and rectangular with a sharp edge, the helve was short and terminated with a short cross-head. The most outstanding feature of this tool was the curved swan-neck stem that joined the blade to the helve. Extra curvature was supplied at this point in order

to keep the labourer's hands clear of the bank when using the spade. This pattern of spade has since disappeared.

The turf spade had a pointed, sharp-edged blade about 12 in. wide and 16 in. long, attached to its handle by means of an iron socket, bent to an angle of approximately 25 degrees. It thus afforded the user control and some comfort whilst using the spade horizontally to cut out thin sections of turf. One side edge of the turf spade was turned up about 2 in. in order to cut the turf with a straight edge. This spade probably originated in the eighteenth century and upon being mass produced was used by all good farmers and gardeners for turf culture. This kind of spade is no longer common, as turf is now raised by machinery.

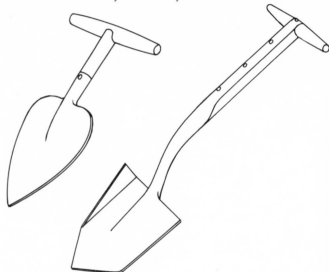

Above: the dung or muck spade
Below: the turf spade

Thigh pads of the kind illustrated were often worn by labourers using a turf-cutting spade. They were looped into the front of his waist belt and hung down in front of both thighs. He rested the cross-head at the end of his spade handle upon the pads and exerted pressure from both legs with comfort.

The frying-pan shovel had a flat, heart-shaped blade which was raised 3 in. high along its rear edge in order to contain loose material or powder. It was furnished with a slightly curved helve to prevent unnecessary bending by the user. A similar pattern of deep shovel is still used in some coal mines and quarries.

The shovel has its helve a little more curved than that of a digging spade. Its square metal head,

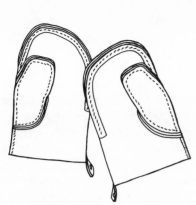

Leather gloves worn by the hedger

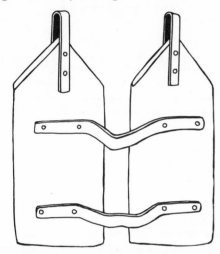

Wooden thigh pads were worn when using the turf spade

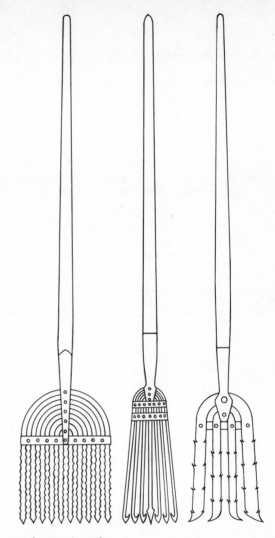

Prongs for spearing eels

turned up along both sides and top edge, is a convenient scoop for taking up fertilizer etc.

Forks have long been used for a variety of purposes in agriculture. The common four-pronged (or grained) variety is still much used by gardeners for digging and raising small crops of potatoes, whilst a three pronged fork easily breaks up tightly packed subsoil. The two- and three-pronged forks for picking up hay, and barley etc. are described elsewhere under the heading **pitchfork** (see page 139).

Eel forks were formed with a wooden pole handle some 6 to 8 ft. long and a head consisting of a number of thin iron prongs, each barbed along its length. A selection of these forks is shown, and it

can easily be appreciated how suited they were for spearing the large and dangerous eels often left stranded when **water meadows** were drained. The long handle was useful when eels were in deep water. They have long since fallen into disuse.

The weeding or docking irons. The operation of 'docking', whereby weeds were pulled from the soil, was a perennial task necessary to keep pasture, hayfield and standing corn from becoming choked. Thistles and docks were most difficult to destroy, since if their deep tap-roots were not completely removed from the soil, new growth would quickly appear. The invention of the horse-drawn **mowing machine** enabled the farmer to retard, if not com-

Frying-pan shovel

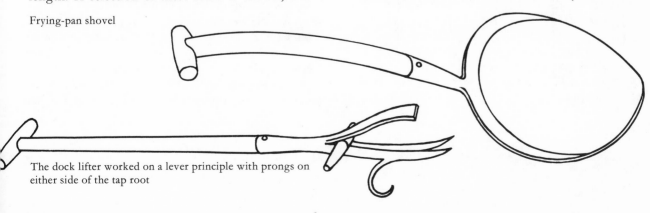

The dock lifter worked on a lever principle with prongs on either side of the tap root

pletely destroy, these weeds, as shearing them off during the winter season with a mower proved very effective. Until this time tap-rooted weeds were extracted individually until the whole field was clear, the labourer being equipped for this purpose with a weeding iron. This instrument consisted of a wooden handle 4 ft. 6 in. long, at the bottom of which was an iron shank 18 in. long and 2 in. in diameter. This shank was flattened towards its end and terminated in a two-pronged claw. A foot tread was provided on the shank which enabled the labourer to push the claw down into the soil up to the point where a projection on the rear of the shank prevented it from going any further than 8 in. deep. By pulling the handle back and using the rear projection as a pivot, he would grip the tap-root between the claws and lever it clear of the soil. Should the dock or thistle break off whilst the labourer was attempting to extirpate it, then he would use a chisel bladed iron, thrust into the ground at an angle, to sever the root and prevent its re-appearance.

The weeding hook is amongst the selection of tools illustrated on this page. It consisted of a light

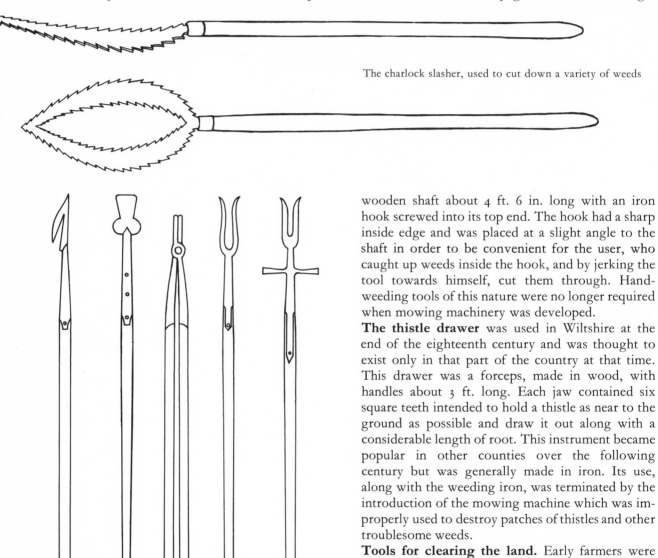

The charlock slasher, used to cut down a variety of weeds

Various weeding tools

wooden shaft about 4 ft. 6 in. long with an iron hook screwed into its top end. The hook had a sharp inside edge and was placed at a slight angle to the shaft in order to be convenient for the user, who caught up weeds inside the hook, and by jerking the tool towards himself, cut them through. Hand-weeding tools of this nature were no longer required when mowing machinery was developed.

The thistle drawer was used in Wiltshire at the end of the eighteenth century and was thought to exist only in that part of the country at that time. This drawer was a forceps, made in wood, with handles about 3 ft. long. Each jaw contained six square teeth intended to hold a thistle as near to the ground as possible and draw it out along with a considerable length of root. This instrument became popular in other counties over the following century but was generally made in iron. Its use, along with the weeding iron, was terminated by the introduction of the mowing machine which was improperly used to destroy patches of thistles and other troublesome weeds.

Tools for clearing the land. Early farmers were forced to expend a considerable amount of time and labour on clearing forests and woodland in order to gain more space for cultivation. The timber was felled with an axe and used for building whilst the

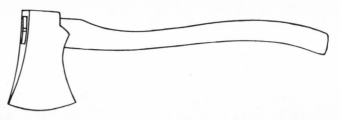

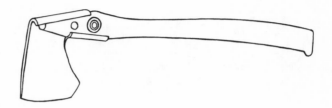

Hedger's axes

branches were destroyed by burning, but the stump and roots remained firmly embedded in the soil. Unless the land was thoroughly cleared it could not be ploughed over, so there was no alternative but to extract the stumps by hand digging. English farmers made themselves a cheap instrument to aid this irksome task, and the idea was no doubt carried abroad to the Colonies where much larger areas of land were being cleared. It was an iron hook about 3 ft. in length with an iron ring passing through a hole near the top, and could be forged for a few shillings. Before it was used, the soil was cleared away from the tree stump and any side roots cut through, then the point of the hook was hammered into the stump and a long lever put through the ring. Two men, one on each end of the lever went round and round and exerted a great pressure on the tap-root which twisted then snapped apart, whereupon the stump could be lifted out and carried away. The invention of the Australian **stump jump plough** in 1876 largely overcame the difficulties caused by tree stumps, roots, rocks and other buried obstacles, but in difficult circumstances, a hole was bored with a long screw-auger, an explosive charge inserted, and the object blasted out of the earth. Land reclamation is now a task for the specialist companies who use heavy machinery to lay forests, whilst the farmer may own or hire a bulldozer for general clearance work. There are flailing machines to deal with scrubland and various tractor attachments, winches etc. for clearing bushes.

Hedging. The hedger can still be seen at work about the countryside, particularly during late autumn and winter when sap is not rising. First of all, the bottom of the hedge, usually one of especially planted hawthorn, is cleaned out and any dead or old branches are removed. Then the new growth is nicked at an angle near the bottom with an axe or bill and the branches are bent over. The hedger is always careful not to cut completely through the

Nineteenth-century billhooks. The lower one is made from an old plough coulter

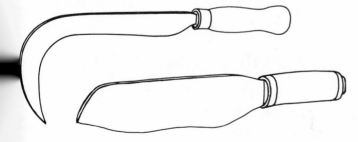

branches or to break them off when bending over, as some portion of the bark must be left intact in order for the top of the hedge to survive. When the branches have been laid, they are woven together and pegged into position with upright stakes placed every six feet along the length of the hedge.

The hedge bill has been used for a multitude of jobs about the farm, but mainly for trimming and laying hedges. They are of various kinds, each consisting of an iron blade set into a short wooden helve, and except for local variation have hardly changed since Roman times. They are intended for severing the slender branches of which a hedge is chiefly composed or if necessary only cutting them half way through so that the branch could be laid over on its side to fill a gap. The hedge worker usually managed with a single bill and an axe which he used for cutting down strong branches. The axe must be included as an important hedge repair tool, and also the chisel-edged cutter attached to the end of a long handle which was used to reach high branches and to prune any trees which grew along the row. Also a long curved blade in the manner of a scimitar was used to trim off twigs and give the hedge its final shaping. Its narrow blade was about 18 in. long and attached at a slight angle to a handle, short enough to be gripped in one hand only. It could be attached by means of a socket to a handle some 4 ft. long and so enable the user to reach the highest parts of any hedge. The hedger now seems to work with an axe, a bill and a saw only. He is unfortunately disappearing from the rural scene and his work is now roughly completed by tractor-driven machinery.

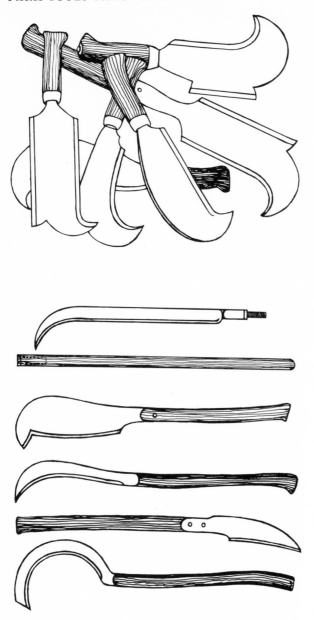

The hedger's tools

A hedge cutting and trimming machine was invented by Messrs R. Hornsby and appeared on the agricultural scene about 1880. It was received with enthusiasm but did not oust the humble hedge-layer, who has continued to employ his craft up to the present day. But extensive trials showed that the machine was capable of cutting and trimming a variety of hedges in a speedy, efficient manner at much reduced cost. It is very likely that the manufacturers took a lead from the **grass mowing machine**, as both appear to have been quite

similar in working principle and general appearance. The hedge cutter was made entirely of iron and moved along on two large road wheels. The motion of these wheels was transmitted through gears to a cutting arm which was composed of two sections, hinged together and extended from the side of the machine directly above the left side road wheel. The outer section was the cutter bar with reciprocating knives which acted upon the hedge, and this was attached to a sliding bar which moved inwards or outwards by means of a crank handle in order to alter the distance of the knives from the machine. The cutting arm was capable of adjustment to any angle and could be raised or lowered to suit the height of the hedge or in accordance with the level of the land along which the machine was moving. It was designed to cut both sides of the hedge from the one side so that the machine could be kept where the ground was most suitable. Two seats were provided, one for the driver who directed the pair of horses, and a second one for the man who attended the cutting arm. It was claimed by the manufacturers that, given suitable conditions, five miles of hedge could be cut in a working day.

The working principle of the grass mowing machine has since been successfully applied to small electrically driven hedge clippers, whilst larger versions not unlike the Hornsby machine have been tried. But modern farmers prefer the speed and power of a large toothed disc, which is set at the end of a flexible arm and attached to the side of a tractor.

The post hole borer, brought into use during the latter half of the nineteenth century was a labour-saving tool specifically designed for landowners, farmers and others whose property required a large amount of fencing. The borer was as illustrated, and operated by one man who pressed it into the ground, turning the cross-handle with both arms so that with each revolution the screw thread forced its way further down. The sharp point was capable of splitting soft stone, and the fragments, along with soil and small boulders, were forced to rise up through the channel of its screw head, and thrown clear at the top of the hole by means of projecting wings attached to the final part of the screw.

Post hole borer

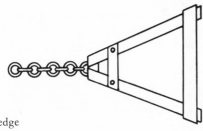

Stone sledge

Mechanical hedge clipper
Left: nineteenth-century post hole digger in open and closed position
Right: twentieth-century post hole digger

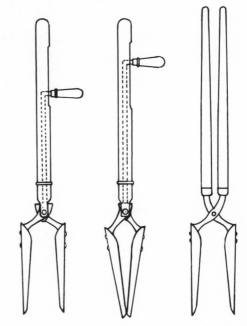

A large number of these drilling instruments were employed in America and Europe for railroad and telegraph development as well as for agriculture.

There are now post hole augers available for attachment to the hydraulic gear of a tractor and automatic drivers to push the posts into the ground.
A post hole digger of American origin was sold in England during the 1880s. This instrument is illustrated and its working action can easily be understood. The merchants claimed it successful in stony, clayey and marshy land, which was proven, and such instruments are still used to a limited extent.
Moles are a considerable hindrance to farmers and although they feed almost entirely on earthworms

67

and not on growing crops, they quickly make a mess of any field if their presence is not checked. The traditional method of destroying moles was to catch them with a steel spring-trap. This entailed locating the run with the aid of a pointed stick, opening with a small spade and then setting the trap in position. Farmers usually paid their men a small sum for every mole's tail they could produce at the end of a week, but the greatest number of killings was made by skilled mole catchers who worked on various farmlands throughout the year. Some supplied the fur trade with mole skins, whilst others recorded their success by hanging the corpses on bushes around the field.

The illustrated instruments were recommended during the eighteenth century wherever moles were numerous. The heavy spiked sledge was about 4 ft. long by 2 ft. wide and was simply drawn back and forth over the casts or fortresses until they were reduced to ground level. The hand instrument was about 4 ft. long and used as a rake to break up a mole's fortress, which can be 1 to 2 ft. high, and 6 to

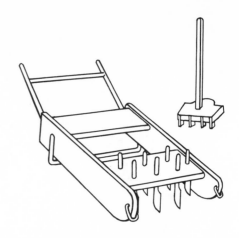

Instruments for breaking down mole hills

8 ft. across. There have long been metal spring-traps for mole catching, but the farmer who deals with a heavy population now uses a specially designed prodding stick to insert earthworms, saturated with strychnine poison, into the deep runs where they are taken as bait.

The Horse Hoe

The horse-drawn hoe. A great deal of the farmer's attention has always been necessarily concentrated upon the control of annual weeds which, left to themselves, would become rampant, shutting out the required light and preventing air and moisture from reaching the roots of growing plants. Until the early eighteenth century, the common means of eradicating weeds was to slash off the tops with a sickle, pull them up individually or hoe by hand between the rows. The latter was the most preventive of these measures and its use also effected some tillage of the soil, but in order to retain a field free of unwanted growth, it was necessary for a large number of labourers to hoe, almost continually, from the first showing of the plants to the time when they were advanced enough to make any further hoeing injurious.

The following account of hand hoeing turnips is extracted from the *Farmers Kalender* for July 1778.

Now you must hand hoe your turnip crop; a work perfectly understood in many parts of the Kingdom; but so much neglected and unknown in others, that it will be proper to enlarge a little on the method of performing it, and in the necessity of the practice. Supposing turnip hoers to be scarce, they demand extravagant prices, or none to be had, order some hoes to be made by your blacksmith; the iron part exactly 12 in. long, and 3 or 4 in. broad, neatly done and sharp: put handles 5 ft. long into them. So provided, take your men into the field, and yourself with a hoe should accompany them: make them hoe the crop boldly, and not be afraid of cutting up too many. Direct them to strike their hoes round every plant they leave, and fix upon the most vigorous and healthy growing ones. By this means they will not be able to leave them less than 14 in. as-

under; for their hoes spreading at every cut 12 in., they cannot spoil your crop by not cutting freely. The work must be done by the day, and you must attend the men well, to see that they cut the land pretty deep, so as to kill all the weeds, and also such turnips as they strike at. In about a fortnight after, send them in again to rectify former omissions, in which time they must break all the land again with their hoes, cut up the remaining weeds, and wherever the turnips were left double, thin them. The men will be awkward in this work the first year, but by degrees they will be able to do it in perfection, by mixing new ones amongst them every year the art will not be lost. The labourers receive payment of four shillings per acre for the first hoeing and two shillings per acre for the second hoeing.

The simple but very effective horse-drawn hoe, introduced in the eighteenth century, could have eased this problem considerably had it been accepted by farmers immediately, but they were sceptical and slow to recognise its value. The horse hoe was not an improved form of any earlier implement but the part result of an entirely new conception of husbandry in the mind of Jethro Tull, a farmer of Shalborne, Berkshire. He also invented the first practical **seed drill** to sow rows of seed evenly and in straight lines along furrows, with the horse-drawn hoe to work between the resulting rows of plants. In 1731 he published his book *New Horse Hoeing Husbandry* in which he outlined his methods and experiments concerning the depth of planting, the quality of the seed, the distance between rows, and the cultivation of various vegetables. In it he also explained the construction of his corn drill and hoe, which he called a hoe plough. He devised his horse hoe essentially to clear away unwanted growth in the spaces between the straight rows of plants sown by his drill, whilst its action aerated the soil and rendered it more suitable for the admission of

Eighteenth century hoe. Pushed by one man and pulled by another between rows of crops

Bushby's horse hoe fitted with broad sharp and revolving
harrow, c.1850

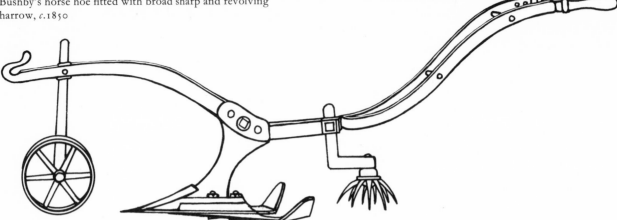

air and rain water. It also procured moisture for the
roots from night-time dews which he considered
to be most valuable in fine weather. He gave the
following instructions for a practical demonstration,
showing how dew moistened the land: 'Dig a hole
in the hard dry ground in the driest weather, as deep
as the plough ought to reach. Beat the earth very
fine and fill the hole therewith; and after a few
nights dew, you will find this fine earth has become
moist at the bottom, and the hard ground around it
continues dry.'

The effectiveness of repeated hoeing was never in
doubt, but his seed drill and his method caused
some consternation at that time, one criticism being
that seed sown in straight rows rather than broadcast by hand was wasteful, in that some soil was left
empty and unproductive. But Tull was convinced
that his method secured healthier, more productive

growth and that the spaces between the rows would
be of greater benefit to the following year's crop
than the customary application of manure.

A great deal of progress was made in the design of
horse hoes before the turn of the century. Tull's
wooden hoe had been equipped with an iron share
and somewhat resembled a plough, but other
gentlemen, Mr Gilbee, Mr Tweed, and Lord
Petres, afterwards provided theirs with a wheel and
various patterns of hoeing tines, of a style which was
to remain characteristic of horse hoes until they fell
into disuse. Whereas they were adequate for coping
with single intervals of a narrow irregular nature, an
inventive clergyman, the Rev James Cooke, was
more ambitious and produced a multiple hoe for
working between a number of rows, providing
they were sown at even distances apart. His machine
was primarily a corn drill, which could be partly dis-

Vipan and Headley's one-row horse hoe for cleaning between
rows of growing crops

mantled, the seed box, cylinder, funnels and coulters all being removed and hoe tines fitted in their place, so that their position corresponded with the intervals between the rows of seed sown previously by the same drill. Before swing steerage was devised for multiple horse hoes it was vital for the driver to hold the horses steady, as any movement to either side would put the hoe tines in amongst the growing plants.

By 1830 there were numerous forms of the horse hoe available and by the middle of the same century the number was considerably greater. Most implement manufacturers had their own versions for working one to four rows of root crops and multiple horse hoes for corn, working six rows or more in the same instance. Almost all of them were provided with a wheel at the fore-end to regulate the depth of work, and handles at the rear for steerage, whilst some were fitted with a harrow trailing at the rear to further remove the weeds. The most effective of hoeing tines were discovered in the early years of the century and were of two distinct patterns which could be used separately or together. A tine with a flat share, made in the shape of a V was employed to work as near to the centre between the rows of root crops as possible, whilst the elbow or L-shaped tines were well adapted to stirring the soil at the side of the corn drills and for working in the immediate vicinity of root crops. The position of the tines could normally be varied to suit the width of the rows, as the collar, by which they were attached to the framework of the hoe, was capable of sliding along a horizontal bar until it was secured by a pinch bolt. The horse hoe was set to work as soon as the young plants afforded a suitable guide for the driver. He walked at the rear of the implement, holding it by the handles and carefully guided the tines along the spaces between the rows, observing with great care that they did not deviate from the correct course and slice through the plants. Only the steadiest of horses were chosen for this work, as any transgression by the horse was more than likely to pull the hoe to one side with it, causing the plants to be extirpated instead of weeds.

The swing steerage hoe overcame the problem of damage caused by the horse stepping out of its proper row, and was used up to the time horse hoes were generally discontinued. One of the earliest patterns of steerage hoe was invented about 1840, by a Northamptonshire man, Mr William Smith. He arranged his hoe as two separate sections, the first section being a carriage frame with wheels, attached by iron rods or chains to the rear section which was a framework with hoeing tines and two handles by which the driver could steer that section only. The flexibility caused by the rods or chains that joined the two sections together allowed the driver of the implement to hold the rear framework and its tines in the intended position, should the horse walk out of the row, and to retain the tines until he brought the horse and the fore carriage section back into correct line. The fore carriage was little more than a pair of horse shafts, supported at the rear above two wheels and their axle. The axle was arched in the middle in order to be clear of the growing plants and it was also divided so that the wheels could be pulled out at either side, to fit conveniently between rows. The rear section of the hoe consisted of an iron bar, made equal in length and placed parallel to the carriage wheel axle, and contained adjustable collars by which the tines were secured. Two steering handles extended from its rear side. Two patterns of tine could be employed with this hoe, one of which was the common pattern with V-shaped share. These were arranged at suitable intervals across the bar for working along ordinary rows, but where the rows were wider the second form was employed. They consisted of left and right hand side tines which were positioned in pairs between the rows. They each curved forwards to a sharp edge where iron wings were attached in order to work horizontally across the spaces between the rows of plants.

The lever horse hoe, introduced towards the end of the nineteenth century, was designed to cope with uneven land or stony soil and could be adjusted to work to the full width of the farmer's own corn drill. Its iron framework was raised upon an axle and two carriage wheels, below which any number up to twenty shares were attached to weighted levers. This arrangement retained the shares below the surface of the soil, but allowed each one to rise up whenever it encountered any obstruction. Steerage of the lever hoe was usually controlled by two handles which projected from the rear of the bar to which the tine levers were attached, and this, in similar manner to the swing steerage hoe, enabled the driver to hold the tines or shares steady, should the horse and carriage frame veer out of line. The shares were raised clear of the soil whenever the implement was turned about, and this was normally effected by a bar that passed

71

beneath the levers and raised them all simultaneously.

Modern tractor hoes have been refined, but operate on exactly the same principle as their horsedrawn counterparts. Many farmers now mount a 'tool bar' at the rear of their tractor and attach a variety of hoe or cultivator tines to it. The tines are adjusted across the bar to suit the width of the rows.

The scuffler was very similar in appearance to the horse-drawn hoe. The difference was in the absence of tines, their position instead being occupied by a large flat share, frequently cast in two or three parts so that it could be adjusted to suit the width between two rows. The broad share was V-shaped, sharpened and serrated along its fore-edge and turned up at the sides. It lay flat upon the ground and penetrated the soil at a shallow depth, cutting off thistles and other weeds that grew between the drills. Small rotary and circular harrows were frequently attached at the rear of the scuffler's frame to fork out all weeds and rubbish after the share had stirred and pulverised the soil. This implement would have been drawn by one horse walking at the fore between the rows of plants, with a driver walking at the rear. They were common until replaced by tractor-drawn hoes.

The shim. This was a heavy implement used in England and Scotland during the late eighteenth century for clearing away bean stubble and other crops from the surface of the land. It was also employed to clear away weeds. The shim was drawn by a team of horses and was built with huge baulks of timber, the purpose of which was to provide strength and weight above a wedge-shaped iron blade which was dragged beneath the implement across the surface of the soil. It was similar in appearance to the wheel ploughs of that time. The front end of its main beam rested on an axle and two wooden wheels, each 3 ft. in diameter, then sloped back for some eight feet to branch into two divisions, each one being about 4 ft. in length. Each branch supported a stanchion for the driver's seat and a stanchion which passed down to either side of the cutting blade. The iron blade was 6 or 7 ft. in length and sharp across its front edge, lying flat upon the surface of the soil and supporting the whole weight of the beams and driver. When the shim was hauled forwards its blade would reduce all weeds and rubbish with which it came into contact, besides levelling the soil. It quickly lost its

popularity when the **cultivator** was invented, although it was retained in Berkshire and neighbouring counties for some time to fulfil the purpose of a **paring plough.** However its dimension was reduced to suit a single horse and its form became even nearer to that of a plough, wherein it contained a beam with two stilts at the rear end and a small wheel or disc-coulter at the fore-end. The blade or share was retained in the previous position, transverse to the line of draught, but it was adjustable through the beam and could be raised or lowered along with the wheel and accordingly regulated to the depth at which the turf was pared away from the soil.

The Cultivator

The cultivator, more commonly known as the grubber, scarifier or scuffler, developed from the **brake harrow** and later superseded it. The cultivator differed from the harrow in that it worked the soil to a greater depth and it could, where appropriate, be used instead of the plough for breaking up whole land. It therefore served to act as part plough and part harrow, but could be employed most effectively where land had to be stirred or cleaned and not turned over or ridged. Its prime function was to clean and stir the soil, bringing up the runners of any weeds, roots or large stones that were buried beneath the surface. The cultivator was not required when farmers cleared their land by bare fallowing every third year, but after about the middle of the eighteenth century this method was gradually laid aside and thus created the need for a strong implement, fitted with deep working teeth or tines.

A definite departure from the harrow towards an implement used specifically for cultivation took place in Scotland about 1784, when several farmers in the area of North Berwick began using a heavy timber framework fitted with seven or eight perpendicular iron tines, each sharpened in order to penetrate the soil. This crude implement was dragged through the soil by horses and oxen. whilst the weight of the framework forced the tines down to a working depth, which could be adjusted by raising and lowering two small wheels at the front of the framework. The term 'grubber' became common in Scotland and was attached to an improved version of this implement when it appeared before the Dalkieth Farming Club in 1811.

The Scottish grubber. The improved Scottish grubber consisted of two rectangular timber frames, one set within the other. The inner frame was equipped with nine cross beams through which were positioned eleven iron tines each of triangular section. The outer frame was carried on four small iron wheels with two handles projecting to the rear for steerage. The tines were given a slight forward angle and each one could be adjusted and then held fast by means of a pin or wedge. This implement was made to different sizes, the largest worked by four horses and one man was equally as effective as the plough when used to stir the winter furrow for spring sowing.

Finlayson's grubber. Finlayson's patent self-cleaning harrow or grubber represents a most important stage in the development of the cultivation implement. John Finlayson, an Ayrshire farmer presented his self-cleaning harrow in 1820, and his design, followed by developments in England, radically changed the appearance of the cultivator, although its function remained basically as before. Its use, however, became extended due to the adoption of various shaped tines, all interchangeable and suited to a variety of uses. Finlayson's implement was the first of its kind to be forged wholly in iron and marked the beginning of practical construction and very effective operation. Only six years after its introduction it received a high degree of fame by successfully breaking up Hyde Park in London. It had nine tines fixed to the cross beams of a low triangular frame, the whole of which was carried on three small wheels. The tines were made to curve upwards from the cross beam to which they were attached before curving down

Finlayson's harrow

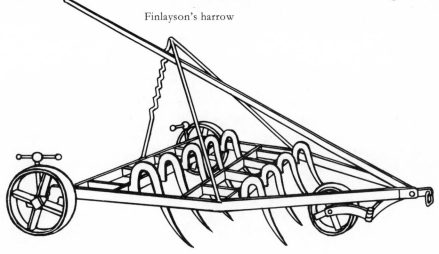

to finish almost parallel to the level of the soil. Finlayson designed his tines in this manner so that the implement could be easily drawn forward and effectively bring all weeds and rubbish to the surface of the soil. The shape of the tines also enabled the implement to remain free from clogging, since any entangled couch-grass or weeds would slide up the curved tines until released. A most important feature incorporated by Finlayson was the device which raised or lowered the whole frame and tines, at intervals of $1\frac{1}{2}$ in., into or out of the soil. This action was effected by the labourer who walked at the rear of the moving implement and moved a long lever which passed from the front single wheel to a measured regulator at the rear. Each movement of the lever simply altered the position of the front wheel and caused the nose of the implement to rise and fall, thus affecting the penetration of the tines into the soil. John Finlayson set the pattern for future development and subsequent improvements were made by various Scottish and English inventors. The two most notable cultivation implements of the 1830s were brought out by James Kirkwood of East Lothian and Arthur Biddell, whose cultivator was manufactured by Ransome's of Ipswich.

Kirkwood's grubber also employed an ingenious mechanism for raising the tine frame and upon this many subsequent ideas were based. This iron implement consisted of two separate parts, having a frame with seven long curving tines and a carriage upon two wheels, the two being connected by the mechanism for raising the tine frame and by rods common to each. The operator, who walked at the rear, guided the implement by means of two handles, and he could easily reach out and adjust the tines in or out of work by means of the screw lever which was connected through mechanisms to the tine frame.

Biddell's scarifier, invented by Arthur Biddell of Playford, attracted the notice of the Royal Agricultural Society of England and was awarded their Gold Medal in 1840. It was skilfully contrived and when drawn by four horses was capable of breaking and stirring eight acres of land per working day. Biddell's scarifier was built of malleable iron with cast wheels and was stronger and heavier than any other cultivator at that time. Two rows of tines were attached to the frame, the weight of which was carried on two large, wide-rimmed wheels preceded by a smaller pair. Chisel points were fixed to the tines and were well adapted for bringing up whole couch-grass and roots without the annoyance of breaking them up. The tines could be removed and those of a different shape substituted as the occasion required.

The following passage gives an indication of practical experience in the use of Biddell's scarifier, being an extract from a letter written by Mr Henry Case, a Suffolk farmer in 1839, to the Secretary of the Royal Agricultural Society of England:

Respecting the use of Biddell's scarifier, and my own application of it, which implement I have used for the last four years in the cultivation of my farm. There is such a variety of circumstances which occasion the scarifier to be used and which determine the different horse power to be applied on each occasion, that my own method of using it on strong hilly land, frequently with four horses, will form no criterion for the guidance of a farmer differently circumstanced. By use of this instrument I can equally well cultivate my farm with twelve per cent less of horses than I could cultivate the same land without it. The land intended for fallow I plough up deeply and as early in October or November as I can, then it lies until the dry weather in March or April, when I scarify it as deep as it has been ploughed, generally three times in a place, each time followed by harrowing and rolling. It will then in most instances be found clean and ready for ploughing. The previous year's fallows to be followed by Spring corn, on my farm are generally scarified with four horses, two in each furrow at length, but on lands of less tenacity than mine this scarifier is used with three horses, two in one furrow, on that side of the stretch where the implement covers half, and one horse in the opposite furrow, in this case a long steelyard **whippletree** is indispensible. I am informed that an admirably constructed caster wheel is made by Messrs Ransomes, which if affixed at the long end of the whippletree, makes it go remarkably well for a single horse. In cleaning my pea and bean stubbles I first use my chisel points and, if the land be very hard go twice over with them, and if necessary then take off the points and affix the broad blades which cut the land clean. At your request I have given this description of some of the uses of Biddell's scarifier, but the practical farmer will vary the uses according to his skill and circumstances, and will require no further direc-

Biddell's scarifier. From the *Farmer's Magazine*.

tions than those in the printed circular. In using this implement I have found it necessary to caution my men against suffering the horses to turn at the ends of the work, without raising the tines from the ground, which is easily performed by means of the lever, and unless they pay particular attention to this, some part of the implement would be likely to be broken.

By about 1860, the cultivator was used extensively in both England and Scotland, and in the spring preparation of land for root crops it had to some degree replaced the plough. It was adaptable to a variety of conditions and would result in a satisfactory seed bed. For autumn cultivation it was used immediately after reaping, to stir up stubble land in order to allow air and moisture to penetrate before the soil was ploughed. Weed growth was best retarded during this season and timely use of

the cultivator would ensure that their seeds were covered and started into growth before ploughing commenced. With adaptation, the implement could be arranged to lift the top layer of soil, without inverting it, so that twitch of couch-grass were almost totally removed. Manufacturers had produced a large variety of models by this time, and although the majority were retained with high reputation for the remainder of that century, tine and share attachments were far from perfect and still required the addition of a heavy framework in order to press them into the soil. It was found that straight or vertical tines pulverised the top layer of soil but tended to pack or compress that which was below its point. They caught up roots, twitch and other buried weeds, but dragged them along instead of drawing them out. Straight tines angled towards the rear of the implement were more effective for reducing surface clods and, whilst they did not penetrate as deeply into the soil as the vertical type, their backwards inclination made a very desirable tilth. The drawback with this tine was its tendency to bury weeds rather than draw them up. Tines curving obliquely forwards into the line of work disrupted the soil to a good depth, moving it upwards and slightly to either side, whilst bringing up whole weeds. Surface clods were not broken down effectively by these particular tines, but were pushed forwards until they fell away. A variety of broad shares, with narrow or wide wings, and with or without points, could be attached to the working end of the tine whenever appropriate.

In 1896 Howard's of Bedford introduced their **'champion' cultivator,** which employed a release action similar to the horse rake. It enabled the driver who was seated upon the implement to raise instantly the whole nine tines out of work in the event of clogging. More important was the addition of a corn sowing box attached at the fore-end of the framework, which deposited six drills of corn as the cultivator moved forward. The sowing mechanism inside the box was activated by chain-drive from one of the rear wheels. About the same time, Massey Harris and Company introduced their Canadian spring-tine cultivator into Great Britain, and compared with those in current use, its draught was remarkably light. This was due to the inclusion of large travelling wheels, and spring-tines which resembled those in the **spring-tine harrow,** but instead of being rigidly secured to the bars of a framework on the cultivator they were hinged in four groups of three along the main wheel axle, and could be raised or lowered by a lever.

The essential difference between cultivators at the beginning of the twentieth century lay in the shape of the framework and the manner in which the tines were attached to the framework. The grubber or Scottish type, descended from Finlayson's and other Scottish implements, retained an almost triangular iron framework with two large wheels at the rear and a smaller wheel at the fore-end. The framework was always parallel to the ground, even when the tines were out of work, being raised and lowered by means of a lever-operated, cranked axle on the rear wheels and a sliding stem on the fore wheel. The tines were situated within the breadth of the framework and between the front and rear wheels, whereas in its opposite type, the English or bar type

cultivator, tines were affixed to a bar which passed across the rear end of the framework and extended out beyond the wheels. This bar was hinged to the framework and raised or lowered the tines into work, the tines normally being made to slide along the bar and into the desired grouping, where they were secured by pinch bolts.

In the early years of the twentieth century, various makes of cultivator of both bar and grubber type were furnished with spring tines, which were made much stronger than those on the harrows. Spring tines were judged to be more effective than the older, rigid form of tine, as their vibrations assisted in pulverising the soil and their elasticity allowed them to ride over obstacles and sustain less damage. Some cultivators were equipped with spring-mounted tines and others with sickle-shaped tines. The former were rigid and attached to the framework or wheel axle by means of spring mountings, whilst the latter were identical to those on the spring-tine harrow, and rotated around the axle to which they were secured. This arrangement altered the depth of tine penetration, and their points could be retained vertically in order to work shallow, or be rotated until they lay horizontal, in the deepest working position.

Cultivators made for tractor attachment were stronger and heavier than the horse-drawn equivalent, and fitted with self-lifting mechanism which could be operated by the tractor driver from his seat. They were commonly furnished with nine tines, designed to work deeper and under greater stress than before, whilst the wheel bearings were also improved and fitted with oil baths to withstand the increased speed. Most of these cultivators were built with two travelling wheels and a grubber-type framework, the fore-end being hitched to the tractor draw-bar, but this arrangement proved unsatisfactory as early draw-bars were not properly designed for heavy loads. To overcome this difficulty, manufacturers attached a small castor wheel or carriage to the fore-end of the cultivator, in order to carry the down-thrust and relieve the draw-bar of its pressure.

A large variety of cultivators are now available, the spring-tine version being perhaps the most widely used of them all. There is also a simple 'tool bar' which is attached to the rear of a tractor and accommodates a row of cultivator tines.

The American corn cultivator. During the first half of the nineteenth century a large number of wheeled cultivators were shipped from England to America, where they were used in addition to the plough for breaking up new land or to prepare fallow land for corn. Horse-drawn hoes also sent out from England were used in the tradition of the older country to eradicate annual weeds that grow between rows of young plants, but they were more essential for producing the loose surface tilth necessary for corn growing in a dry climate.

The practical development of a two-row corn planter commenced in 1853, and with it came a demand for a more suitable form of cultivation implement, one that was more effective and faster than the imported but well used English pattern of horse hoe and cultivator. Out of this necessity emerged the 'straddle-row' corn cultivator, which was immediately adopted together with the new corn planter in the Western States of America. The combination was successful and when used together the two implements enabled any farmer working unaided vastly to increase his amount of growing corn without neglecting other tasks about the farm. The first of many patents towards an American pattern of cultivator was granted in 1856 to George Esterly, a Wisconsin inventor who later became famous for his harvesting machines. His new implement resembled neither the English horse hoe nor the English cultivator in appearance, yet it served as both. He attached a straight axle with two wheels at the rear end of a draught pole, the other end of which went forward between the horses. Two 'shovel' hoes were attached to an iron bar which was 2 ft. behind and parallel to the wheel axle, the two being loosely coupled together with chains. The bar was provided with two handles which extended at the rear and enabled the driver to retain the hoes in their working position should the horse and wheels move out of correct line. The handles also allowed the driver to lift both hoes out of the soil whenever an obstruction was encountered or the implement was turned about.

In 1867 P. Conrad of Illinois patented a straddle row sulky cultivator in which the wheel axle was arched at the centre where it was joined to the draught pole. His patent also covered a sleeve coupling by which the two separate gangs of hoes were attached to the axle and allowed either gang to be raised out of work by means of a handle. Conrad's implement, as illustrated, represents the general appearance of the American corn cultivator before about 1880. Following that date, patents were

taken out by Gilpin Moore, E. A. Wright of Iowa and others, for the use of single and double acting springs which exerted a lifting strain on the gangs after they had been raised above the operating position and alternately they bore down on the gangs when they were working in soil. The long draught pole remained a popular style for many years, as it provided the gangs with a long easy swing which suited many American farmers. However some manufacturers omitted the beam and compacted the implement by attaching the horses directly to the wheel axle; the later found more favour in the long run, as did those implements which continued to include a seat for the driver. Walking cultivators, where the driver walks, became popular for a short time towards the end of the century on account of their light draught and inmanoeuvrability. With this version, the farmer walked at the rear holding the gangs in working position between the rows instead of being seated upon it. By the end of the century, the cultivator had been modified so as to be appropriate for uses other than corn. Spring-tines, similar in kind to those invented by David Garver for his harrow, and sharp-edge discs, both plain and cut away, were included and found favour in place of the traditional shovel hoes.

The Harrow

The first form of harrow to be used by man was most likely a branch taken from a thorn tree. Sometime later the thorns were replaced by wooden spikes fixed into the underside of a forked branch with logs tied across it to exert additional weight upon the spikes. This would have loosened the soil more effectively than a thorn bush. The ancient Japanese may have used a harrow with sharp-edged stones arranged side by side in a wooden framework, and the ancient Egyptians a bush harrow made with dried palm leaves. The Romans almost certainly used a harrow in the form of a plank or log fitted with spikes on the underside. They employed it to clean out weeds and roots in freshly ploughed land, and then they reduced the soil to a fine tilth by dragging a wicker hurdle across it. The latter may have had wood or iron teeth driven into it, but more likely the branches used for its construction were left with a large number of hard spiky off-shoots which would perform equally well. The harrows used in Europe about the turn of the tenth century would probably have appeared as low, square or rectangular, open-framed platforms, consisting of wooden beams supported on iron or wooden tines. When they were drawn straight forward by oxen little control could have been asserted over them, as they would have tended to swing out of line and climb out of the soil unless weights, perhaps in the form of logs, were placed on the top side. Gradual improvement took place and by the sixteenth century a triangular harrow had appeared, the reason being that such a shape allowed the tines to be 'staggered' across the framework, rather than set in regular rows, so that each one did not then follow in the track of the one before it. At the same time farmers began to use **wooden rollers** and **clodding mells** to smash down the largest clods.

The bush harrow as used up to this present century, closely resembled the thorny bush which the earliest farmers dragged along by the roots. Its simplicity of form and construction ensured popularity in many countries up to the end of the nineteenth century, and as late as the present day in remote regions of the world where modern implements are yet scarce. During the seventeenth century, Flemish farmers may have been responsible for an improved form of bush harrow which subsequently spread to other European countries, including England, over the next hundred years. It

was recommended for scratching up the surface of grassland, as a means of dispersing decayed matter over the same surface, and for thoroughly effecting a fine tilth. It usually consisted of a flat timber framework, formed by three or four bars about 8 ft. in length, secured about 1 ft. apart in parallel by short bars across each end and strengthened by diagonal braces. Thorny branches were woven between the bars, the main body of which projected out at the rear of the frame to form a crude but reliable harrow. Most farmers constructed their own version, with materials at hand. Some used old farmyard gates interlaced with branches and others used wickerwork hurdles with twigs inserted into the weave. After about 1840 the various forms of bush harrow were gradually replaced by a single pattern of **web harrow**, made in iron, that achieved better results, since it scratched the surface to a greater depth and gave the seeds, lying upon it, an even covering.

The brake harrow was the strongest form of harrow. Originating in the Middle Ages, it was similar to, but much larger in dimension than, the common harrow, and similarly varied in shape from one locality to another, being sometimes triangular, square, rectangular or rhomboidal. Its exceptional weight, achieved by using huge timbers for the framework and square irons for the tines, was utilised during the winter for the purpose of breaking up obdurate, clayey land that would otherwise defeat the common harrow. Early in the nineteenth century the brake harrow was replaced by the first forms of **cultivator,** an implement that rapidly found favour owing to its ability to work at a greater, more consistent depth.

The levelling harrow was similar in its function to the **brake harrow** except that it extended the power to level the land quite evenly after having pulverised it. This implement embodies two separate frames, each with tines, one triangular frame immediately behind the horses and the other rectangular, which trailed at the rear. The latter was hinged to the rear of the triangular frame and was equipped with handles to raise or lower it into work. When drawn forward, the triangular frame remained horizontal to the ground, and with its tines reduced the clay or soil to small pieces. The rectangular frame which carried with it a greater number of tines could at any time be lowered in order to remove soil from high parts of the land, carry it forward and thus level out any inequalities. By using this implement the ground was reduced to a plane surface without depressions or elevations.

The eighteenth-century harrow. During the eighteenth century the common harrow consisted of one, two or three separate but identical sections, termed leaves, placed side by side and chained behind a long draw bar, which in turn was connected to animals. Each leaf normally had four longitudinal wooden beams, known as bulls, which were joined together with the same number of cross bars. The timbers were cut about 4 ft. in length and 3 in. square for the bulls, and about 4 ft. in length and 2 in. square for the cross bars. Five or more iron tines were fixed through every bull and held securely in position by being driven through a hole smaller in diameter than itself.

Grey's wet weather harrow, an early nineteenth-century implement, consisted of four or more leaves attached to the rear of a beam which was carried at either end on a wheeled carriage, each carriage having shafts to a horse. It was intended as an aid to the farmer sowing wheat on wet ground, but the chain-link harrow proved more successful in the long run.

The rhomboidal harrow. The transition from rectangular to rhomboidal harrows was completed

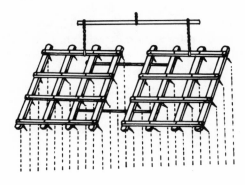

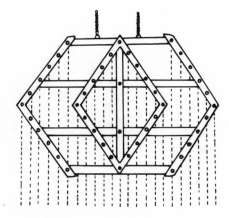

Early nineteenth-century harrow
Top: rhomboidal harrows
Bottom: angular-sided harrows

by the early years of the nineteenth century and the latter was considered, at that time, by many agriculturalists to be the ultimate form of this implement. John Loudon described the Berwickshire harrow as the best implement of this kind and he mentioned a lighter, smaller version suitable for covering grass seed. The rhomboidal harrow retained the constructional details and requirements of the former rectangular harrow, varying only in shape. The object of this difference was to attain a more controllable, uniform action of the implement whilst working to a greater width of surface. In this harrow the rhomboidal-shaped leaves were not allowed to travel freely but were held together, in the necessary position, by the use of two coupling hinges, one near to the front end of the frame and the second one near to the rear. Each leaf was attached to a draw bar by a hook and chain. When the harrow was drawn forward the cross bars remained at right angles to the line of draught, whilst the bulls were at a constant angle to this line. The tines, being equally spaced out along the length of the bulls, were thus caused to indent the surface at regular intervals with no two tines following in the same track. The same principle of design is applied to some modern harrows.

The drag harrow was used in many southern districts of England between the middle of the eighteenth and nineteenth centuries, for clearing land and preparing it for seeds. It was generally triangular in shape, the sides being 6 to 7 ft. in length and fitted with a number of flat 'duck footed' iron shares. In the absence of wheels, the whole weight of the wooden triangle was carried by the shares. A similar implement known in Devonshire as 'the Tormentor' moved along on three wheels. The sheer weight of both Tormentor and drag necessitated a number of horses to draw them along, and both became obsolete as **cultivators** and harrows were improved. Some circular harrows were produced during the early nineteenth century, but they were tried and found not so efficient as other types.

The grass seed harrow was used lightly to cover grass and clover seed with soil. Although identical to the common harrow in structure, shape and number of tines, the noticeable difference was its small dimension, the bulls being only some 3 ft. in length. Since grass seed required one light harrowing so that it would not be buried too deeply, wings or part-harrows were attached to either side of the

79

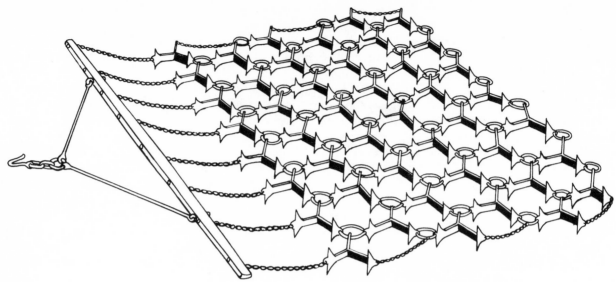

Iron link harrow

implement in order to cover as wide an area as possible, but the extreme corners of the wings were cut away and allowed neat and speedy turning. Lightweight harrows of this kind are still very important to the farmer.

The web or chain link harrow was an alternative implement for covering over grass seed, developed during the 1840s by the hollow drainage advocate, Mr Smith of Deanston. Later in the century his design was exploited by a variety of implement manufacturers and underwent a series of minor improvements, but it remains to this present day in an almost identical form to the one first conceived by Smith. To form this harrow, a large number of iron discs each 4 in. in diameter, 1 in. thick at the centre and tapered towards a sharp serrated edge, were arranged at regular intervals within an iron chain web. The discs were caused to rotate when the whole pliable iron blanket, some 6 ft. by 4 ft. was dragged over the soil, and their rolling action tore and abraded the land surface, smashed clods, and disturbed soil to a depth sufficient to cover small seeds. Its operation was greatly superior to that of any other seed harrow. Most modern farmers own and use a web harrow or some deviation of it.

The traineau. On his return from the Netherlands in the 1830s, the Rev W. L. Rham, Vicar of Wink-field, Berks, said: 'If a door were taken off its hinges and dragged over a field it would become a traineau.' The implement was essentially Belgian and had been employed in that country for a century or more to level the surface of light soil. It

was not a substitute for the harrow, since a tri-angular form of harrow with wooden tines was being used in Belgium at that time. It was more of an alternative to the stone roller, though it did not compress the soil to the extent of the roller but merely flattened the top surface with a grinding action when it was dragged along. The traineau consisted of a stout frame made to the shape of a trapezium, the widest end of which was boarded across with stout flat timbers in contact with the ground, whilst the end nearest to the horses was left open. It was weighted with stones to suit the condition of the land, and in some districts of Bel-

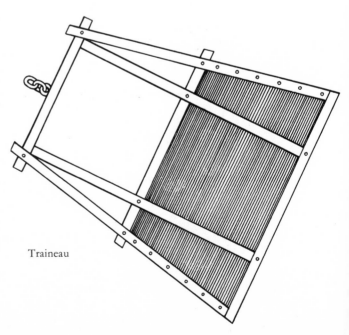

Traineau

gium a large hurdle, woven from willow branches, was dragged flat along the ground to fulfil the same purpose. The reverend gentleman reported that, despite the crudity of these devices, they produced a very fine surface on dry land.

A similar device came to be used in England later in the same century and was commonly called a plank-drag or rubber. It was normally improvised by the farmer himself, but a wooden version light enough to be drawn by one horse could be purchased if necessary. It was usually 6 ft. or slightly more in width, by about 4 ft. in length and the underside was made rough by overlapping the planks used for construction. The edges and corners were reinforced with iron strip and it was connected to the horse by two chains from the fore-edge. A heavier version was often formed by fastening a number of iron bars together with short lengths of chain. This type of implement was laid aside about 1930, and replaced by tractor-drawn harrows.

The Norwegian or revolving harrow, so-called but made in both England and Scotland during the early nineteenth century, was not specifically a harrow, but none the less was capable of performing as such. Morton's revolving harrow was one of the first implements in this class. It consisted of a low iron framework carrying three parallel axles set about 12 in. apart, on each of which were a number of star-shaped wheels, each some 13 in. in diameter and sharpened at the points. The framework was contained inside another, which travelled along upon four wheels, and by the use of a lever arrangement, either frame could be raised or lowered so that the whole machine could travel on the wheels only, or on the star points only. When the harrow was dragged forward, the star wheels were forced to revolve and the action of their points effectively reduced the size of all large clods and at the same time penetrated to a considerable depth into the earth, tearing the surface to pieces and producing tilth along a 4 ft. wide strip. Its pulverising effect was improved by James Kirkwood of Tranent towards the middle of the nineteenth century. He re-

designed the star wheels and applied a better mechanism for raising or lowering them into the work. But his improvements were somewhat too late as this rotary harrow was eclipsed by the new web harrow, particularly suited for preparing a fine tilth, and the excellent **Crosskill clod crusher.**

The snow harrow originated in Scandinavia and during the nineteenth century came to be manufactured in Scotland. It was an essential implement only where a prolonged harsh winter was common and was employed by farmers in such areas to free their land from the grip of ice and snow. Its function, when pulled along, was to turn up and expose frozen snow and to cut thin ice into strips or lumps suitable in size for collection by a scraper. A single seat was provided at the immediate centre of the bull upon which the driver sat and controlled the one horse. If necessary extra weight or stability could be provided by a second man holding the single handle that protruded from the rear of the implement. A similar, but mechanical implement is still employed for ice breaking in colder countries.

The drill harrow was designed particularly for harrowing amongst ridged crops and proved to be so excellent when used with potato plants that it consequently became known as the potato, or saddle-back, harrow. It moved the whole of the surface and was particularly effective for destroying annual weeds which became established before the potatoes sprouted through the soil. This harrow was comprised of two identical leaves, each one arched to the extent of 5 in. at the centre to accommodate the curvature of a potato ridge. Two leaves were linked together by adjustable bars, the variance of which arranged each leaf to fit over adjacent ridges of potatoes with a single horse walking between them. Each leaf was comprised of three bulls, 3 ft. in length, across which were positioned the three arched cross bars, each about 2 ft. 6 in. in length. The 5 in. tines were screwed through both bulls and cross bars. One, two and three-row potato harrows have been made by various manufacturers since the beginning of the nineteenth

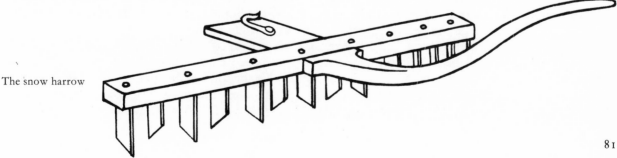

The snow harrow

century, and they are still occasionally used at the rear of a tractor, the tractor wheels being set to run in the spaces between the rows.

The zig-zag harrow first came into widespread use about the middle of the nineteenth century and after being modified only slightly, is employed to excellent effect in modern farms. From the time of its invention this harrow has been manufactured entirely in iron and the use of this material provided a low, compact implement which did not swing out of line as other various shaped harrows were inclined to do when drawn forward. The leaves were identical in shape and size, made to an angular shape resembling the letter Z. The complete zig-zag harrow was formed by the chaining together side by side of three leaves, each one being attached to a long draw-bar. Each leaf was supported on about twenty tines which were screwed through the bulls and held fast by a retaining nut. The tines were tapered to a sharp or rounded end, according to requirement. For most tasks, a set of three leaves together was adequate, but if necessary three more could be chained to the rear and the harrow then occupied an area about 12 ft. square.

Seeding harrows were very small in comparison to other forms of the implement. They were generally no more than 18 in. in width, and about the same in length; an ideal size for trailing behind a seed barrow etc., in order to cover the seed. They were made in wood with iron tines, or wholly in iron to a square or triangular shape. They were gradually laid aside as seed drilling mechanisms were perfected.

Disc harrows were primarily developed in America. In 1847 a United States' patent was taken out for a single steel disc with a sharpened edge, to be attached by means of an iron bracket to the side of a plough. The function of this disc was to slice or cut through soil as it was turned over by the mouldboard. It could hardly be termed a harrow in the tradition of that implement, but it did mark the beginning of a development that came to fruition before the end of the century, and provided a pattern of harrow which is constantly used on the land today. The ancient Japanese are believed to have employed a disc harrow in which sharpened discs were placed side by side at intervals along the length of an axle, with sharp vertical blades following at the rear to pulverise the soil still further. This idea emerged and was patented in America during 1846. It consisted of sharpened steel discs arranged along an axle, and at that time was not intended as

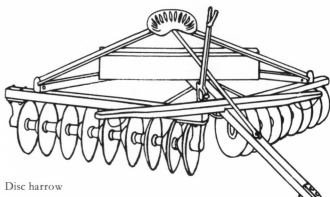

Disc harrow

an independent implement but was made for attachment at the rear of a seed drill in order to break down the soil and cover the dropped seeds. In 1854 another American inventor, Mr H. Johnson, obtained a patent for a harrow which involved two separate rows of discs for the purpose of tilling soil, and it was from about this time that manufacturers became interested in the idea. Improvements were quickly forthcoming and by the 1880s many different patterns of disc harrow were on sale. By the turn of this present century it had found its final form, being then manufactured and used in those countries where the conditions were suitable. It comprised a number of sharp-edged concave discs, between 12 and 20 in. in diameter, mounted 6 in. apart along two separate axles which were set side by side at a slight angle to each other. The effects of the discs when arranged in this manner was first to slice through the soil and then to throw it sideways in a similar manner to the mouldboard of a plough. Each disc was provided with a scraper to remove any soil that would otherwise adhere inside the concave, and these were put into and out of contact through the movement of a lever. In some models the two sections could be set at different angles or were arranged so that one section travelled in front of the other, with the two overlapping slightly. A seat was provided at the centre of the harrow for a driver who controlled the horses and made separate adjustments to each section as occasion would demand. This form of implement was readily adapted for tractor haulage, and often two complete harrows were drawn in tandem, having the concave discs set in opposite directions so that the soil was cast to one side by the first gang, and then returned to its place by the second. A cutaway concave was sometimes employed on the rear gang to pulverise the soil still further. The disc harrow, changed little from its early tractor-drawn form, is still extensively used in many countries.

A spring-tine harrow was invented by David L. Garver of Hart, Michigan, and patented in 1869. He arranged sixteen steel tines in four rows across the beams of a low, 6 ft. square wooden framework. Each tine was secured by one end to its beam and curved upwards, then over to the rear, like a sickle, before tapering to a sharp point for scratching the surface of the soil. The weight of the framework prevented the tines from coiling under to their natural position and ensured that the points of the tines were under tension against the soil, whilst at the same time they were able to ride up over any obstruction, and spring back into place. It was eight years before the inventor could interest a manufacturer in his idea. When it was finally brought onto the market in 1877, it was soon put to useful work by farmers in the central and eastern States of America. An American make of spring-tine harrow was demonstrated in England at the Birmingham Royal Show in 1898, and by that time lever arrangements had been added in order to adjust the pitch and working depth of the tines. The response was favourable, and in a short time licence was granted for it to be manufactured in Britain, where in later years, as in America, larger versions with thirty or more tines were developed for tractor haulage. This pattern of harrow is extremely important in modern farming.

The mouldbaert. A mouldbaert of some form was used in Holland for a century or more before British implement manufacturers brought out their own particular versions in the early nineteenth century. They called it the **levelling box.** The mouldbaert was primarily a Dutch implement used for levelling newly ploughed soil and to reduce other irregularities about the land. In reality it was a huge shovel, capable of holding 500 cwt. or more of earth, drawn by two horses and expertly manoeuvred by a single man. It was constructed mainly of timber, 4 ft. in length and 4 ft. wide, with a concave base. The underside was reinforced with plate iron to withstand constant friction, and the iron also projected out beyond the front edge to form a sharp lip. The haulage attachment consisted of two chains fixed by a ring and hook to either side of the shovel and then brought together some 4 ft. forward at a **whippletree.** A single stout handle, 6 ft. in length, projected from the rear end of the shovel and to the end of this handle was fixed a rope some 14 ft. in

length. The driver grasped the free end of this rope in the same hand that he used to retain the handle. Whilst the mouldbaert was stationary he would pull down the handle in order to raise the free edge of the shovel. He would then start off the horses with the mouldbaert sliding along on its base. When he approached the heap or blemish that had to be removed, the driver would raise the handle to the full extent of his uplifted arm and cause the free edge of the shovel to come down into the base of the heap. It would rapidly gather up earth as it moved along and when sufficient had been taken up the handle was depressed. The earth then fell back into the shovel for conveyance to hollow ground. The driver would then throw up the handle, causing the fore-edge to catch the ground, upon which the whole shovel was tipped fully forwards until the handle was retained by the whippletree. By this means the earth was left behind and flattened by the edge of the shovel as it passed over. The final movement in the operation was caused by the driver pulling on the rope which reversed the shovel back to a suitable position to receive another load. The English adaptation was equal to the mouldbaert in its construction and capacity, but it is doubtful if its drivers would have operated with the expertise of the Dutch. Long experience with an implement of this kind enabled them to work without halting the horses and to level the earth as it was deposited. American inventors made many attempts during the nineteenth century to perfect a mechanical shovel, and these eventually resulted in modern earth-moving equipment such as the bulldozer.

Top: nineteenth-century mouldbaert
Bottom: twentieth-century earth scoop

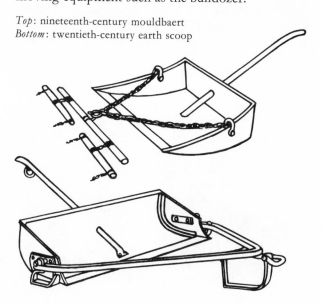

Rollers and Clod Crushers

The functions of the roller in cultivation have been numerous. It was first intended as a means of producing a firm, compact bed for seeds and subsequent plant growth by breaking down large clods of earth left behind by the plough or harrow, and on some occasions it was needed to compress the soil over newly sown seeds or to consolidate and prevent loss of moisture on peat and grassland. The use of a roller in English agriculture was first observed in the late sixteenth century, and a century later a roller was invented to distribute manure over the surface of the land. Rollers were gradually accepted by eighteenth-century agriculturers as an alternative to the tedious task of smashing down individual clods of earth by hand with the **clodding mell,** and by the end of that century the clodding mell had been entirely replaced by a variety of different rollers mostly constructed in wood, iron or stone. Some rollers were fitted with spikes and others involved wood and iron discs placed alternately along an axle. All of the rollers were of different length and diameter, the dimensions of which were in dispute at that time. It was however generally accepted that a roller should have as small a diameter as possible, whatever its material of construction, since the larger the diameter of the roller the less the resulting pressure would be due to it resting too much of its own surface on that of the soil. The material used for constructing the roller was largely responsible for its eventual success or failure. Wood quickly wore down and was too light unless the roller was made either with a huge impractical circumference or was supplied with extra weights such as rocks attached to its frame. Stone was suitable as regards weight but was very likely to split. When it became readily available, cast-iron proved to be the most appropriate and manufacturers proceeded to use this material for their rollers which rarely exceeded 2 ft. in diameter, the total weight being decided by the thickness of the iron that comprised the shell of the cylinder.

The common roller. Before cast-iron became the accepted material for its construction the roller was a simple cylinder of wood or stone, provided with an axle. Early agricultural rollers varied between 15 and 30 in. in their diameter and were normally about 6 ft. in length, except those employed for flattening ridges which were made to smaller dimensions. Shafts for two horses and a scraping device to keep the roller clean were essential. The shafts extended forward from a heavy timber frame that contained the roller, and it was common practice to position a large wooden box above the frame in which to place any extra weight thought necessary and also to retain any odd stones found in the field. Cast-iron rollers, with the correct weight in relation to the diameter, became plentiful early in the nineteenth century. The necessity of carrying extra weights was then dispensed with but, in some instances, the box was retained and made watertight so that liquid manure could be carried with the roller for application to spring crops. The liquid was released through numerous tiny holes pierced towards the base of the rear side of the box, and applied in a continuous but regulated stream as the roller progressed.

The divided cast-iron roller was made similar in its size and appearance to the common land roller, but it was divided exactly half-way along its length into two identical cylinders, each of which could rotate independently of the other. The greatest advantage of having the roller constructed in two sections was the ease it afforded whilst being turned. The division enabled this roller to be manoeuvred right about, by means of one cylinder going forward whilst the other revolved backwards. This prevented the soil and its crop from being pushed up, which was the result of turning one long cylinder. Either cylinder could be replaced should it become worn or damaged, and rollers of this kind are still an asset to the farmer, nowadays being hitched in tandem to a tractor.

The spiked roller. A roller with iron spikes around its entire circumference was used in England about the middle of the eighteenth century. It is likely that a similar roller had appeared before it, but with wooden spikes rather than iron ones. The inventor of the iron spiked roller remains obscure, but it is probable that he intended it as an alternative to the harrow. In order to prepare a suitable seed bed from stiff, heavy clay, the larger lumps had first to be broken down and the crushing action of iron spikes as the roller was pulled forward achieved this more effectively than the plain common roller. The spiked roller was fashioned from the trunk of an elm tree, having a finished diameter of about 18 in. Iron spikes, 7 in. long, were driven into the wood to half their length, at intervals of 2 in. over the entire breadth of the cylinder. During the nineteenth century, improved versions of the harrow and the cultivator replaced this roller but a similar implement was still used in the south west of England for the tillage of wet land. This roller consisted of twelve or more thin, sharp-edged iron discs, 18 in. in diameter, that were interspaced at intervals of 4 in. with circular pieces of wood made to half the diameter of the iron discs. Both were threaded along the same axle. Weights were applied to the carriage frame of this roller which forced the iron discs to penetrate down into the soil until they were prevented from going deeper by the wooden discs. With this implement it was possible to incise parallel gutters over a large area of wet land, a task that would necessitate some form of self-clearing device being attached to the roller.

The hand turnip roller was an essential instrument before rollers of this type were attached to the **turnip seed drill,** whereupon the two separate operations of pressing down the ridge and dropping the seed were completed in the same instance. It continued as an adjunct to the turnip seed drill where a roller was absent, in order to compact the soil in readiness for the seeds and also to preserve the moisture content of the soil. It consisted of two concave cast-iron rollers, each about 18 in. in length, 14 in. diameter at either end and 8 in. diameter in the middle. They were mounted upon the same axle and contained beneath an iron framework from which ropes or wooden handles extended forwards. The position of the rollers upon the axle seldom needed any alteration, as two rollers of that dimension placed side by side were convenient to travel along the tops of adjacent ridges. Such rollers are still employed for turnip culture, but they are attached at the rear of the drill.

The drill roller was particularly used in the eastern part of England during the late eighteenth century for making narrow channels or drills some 4 to 5 in. apart in newly ploughed soil. Corn seed was then scattered hand broadcast and covered by bush harrowing. The early form of this roller was probably a heavy wooden cylinder, with protruding bands of iron to make the necessary channels. Later, like most other agricultural implements, it was constructed of iron, in this case cast-iron rings spaced out along an axle.

The furrow pressing roller, as patented by Robert Berriman in 1806, was largely an adaption of the **drill roller.** Its purpose was to follow behind a plough in autumn and consolidate newly turned land prior to sowing wheat. By operating the roller, two parallel gulleys of suitable depth to receive the seed were pressed into the soil, and the resulting firmness in the seed bed assisted the growing plants during the winter and duly prevented the ravages of wireworm. The implement was as illustrated, each wheel being about 2 ft. 10 in. in diameter and $5\frac{1}{2}$ in. in breadth, with V-shaped rims. In a short time this implement had been modified and more pressing wheels were added so that a greater number of drills could be formed at the same time. By the middle of the nineteenth century it was commonplace for six or more wheels to be arranged at intervals of approximately 10 in. along one axle. The early twentieth-century furrow pressing roller closely resembled Berriman's original design but by that time they were fitted with turn-about shafts, which enabled the farmer to reverse his horses at the headlands without the trouble of turning his implement. Improved seed drilling equipment gradually rendered the furrow pressing roller obsolete.

The Crosskill clod crusher, as patented by W. Crosskill in 1841, represented a considerable advance in the crushing capability of the roller at that time. His implement could also be used as an alternative to the **grass harrow** on light soil which had been sown with seed. Crosskill's roller was immediately recognised for the efficient manner in which it broke down clods of earth. It was formed by loosely fitting a number of cast-iron discs, each 30 in. in diameter, along the length of a 5 ft. 6 in. horizontal axle. Each disc was serrated around its entire circumference like a large cog wheel, with additional wedge-shaped radial teeth on both of its sides. As the roller was hauled forward by horses, the discs revolved and large clods of earth were broken down by the insertion of the teeth, which caused the surface of the land to be sufficiently pulverised. A pair of carriage wheels were sold with this implement so that it could be moved from the farmyard to the field without causing damage to the road surface. The roller axle was made to project out from either side for fixture of the wheels, and this could only be achieved by digging holes in the earth beneath their position. The implement with wheels attached was then hauled forward out of the holes.

The Cambridge roller. Mr W. C. Cambridge patented a clod crushing roller in 1844 in which alternate plain and toothed iron discs were loosely placed along a horizontal axle. This roller was manufactured in a variety of widths between 5 and 10 ft. and cost between £10 and £20 according to size. They were equipped with a heavy wooden framework on which extra weight could be placed if required, and also two long shafts for the horses. From his experience with this implement the inventor developed and patented an improved disc roller that remained popular until this present century. The cast-iron discs of which this roller was comprised differed in their diameter, being 18½ in. and 19½ in. alternately. All were threaded onto the same axle. Each disc had a 3 in. projecting ridge around its whole circumference and the axle holes were placed off-centre in order to throw the discs out of line as they revolved. By means of this arrangement he achieved a peculiar motion that was very efficient in its effect upon the soil and at the same time the roller was self-cleaning. The name of the inventor was applied to a succession of rollers, similar in type to his original patent, and similar patterns are still used.

The furrow roller was devised for rolling the furrows in mountainous and hilly districts where other rollers could not be employed. It was invented by a Mr Pinchard probably during the late eighteenth century and was commonly called the bellying roller. Made in cast-iron, about 4 ft. in length and 2 ft. in diameter at the centre, it tapered to a near point at either end.

A scorching machine for charring the soil after harvest and for burning weeds, straw and bean stubble was invented by Messrs Jones and Draper of Chorlbury. It attracted the observation of the English Agricultural Society in 1840. Few details of this machine remain, but it would seem that it comprised a fire box mounted on a frame some 3 ft. in breadth. It incorporated a fan, turned by means of a chain and gearing from the carriage wheels which blew the flames down through apertures in the base of the frame. This type of machine was rare and was never fully accepted on the farm, due to the hazards that fire presented. However, the idea has been revived to some extent during the past decade on the lines of a military flame thrower. It is most suited to clearing debris and diseased crops from the surface of fields, but is not practical for controlling weeds between rows of growing crops as they too are likely to be scorched.

A thistle scythe was invented in the first half of the nineteenth century for the purpose of cutting down thistles and other troublesome weeds in a more efficient manner than could be achieved by the common scythe or the **thistle drawer.** Even if this implement had proved to be effective it is doubtful whether it would have achieved much popularity owing to its large dimension and sizeable area of cutting edge, which had to be kept sharp and undamaged. It consisted of a timber framework, in the shape of an Isosceles triangle that lay flat upon the ground, the two equal sides being 21 ft. long and the base 15 ft. The triangle was comprised of a number of square and triangular sections, each constructed in timber and loosely linked together by iron hooks and rings, so that when it was drawn forward by a chain attached to the apex, the whole arrangement could assume the curves and ridges of the land over which it was travelling. The leading point of the implement was furnished with a sharp-edged share, whilst the sides of the triangle had scythe blades attached for the purpose of slicing off thistles a few inches above the root. When it was not being used, either side of the framework could be folded in towards the centre in order to protect the blades and reduce the bulk of its area. When drawn by two horses and attended by one man it was estimated to reduce twenty acres of thistle-infested land in one day. This implement did not achieve popularity on any noticeable scale.

Early nineteenth-century harrow

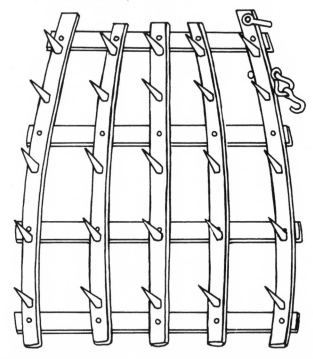

The following is an article written by Mr W. Crosskill, the implement manufacturer of Beverley, Yorkshire, for the *R.A.S.E. Journal* in 1840 (Vol. 2):

As the cost and wear and tear of agricultural implements have now become a serious consideration in the farmer's outlay, and as other improved implements, both of local and general character, may be expected shortly to be brought into use, it must surely be deemed a point of importance to render this item of expense as little burdensome as possible. In order to accomplish this I would advise farmers to bear jointly the expense of such implements as are only required for particular seasons: but more especially to contrive by care and good management to make the implements as durable as possible. To effect this, might not every farmer have a suitable shed, that would admit light and air, with a clean hard floor, the walls being whitewashed every year; within which building each implement should have its proper place?

When the ploughs are done with, let them be washed and put in their proper places; let the same be done with the drill, and so on with all the machines on the farm. The cost of this will be trifling, compared with the advantage. In order to effect it, select the most likely agricultural labourer on the farm; put the implements under his care; make it a strict rule with all the men that each implement done with for the season shall be brought to one particular place, say near the pond or pump; the man having charge of the implements must then wash and clean them well before putting them into the shed, and at a convenient time, when not otherwise engaged, or in weather when outdoor work cannot be performed, get them repaired and paint them. At the end of this shed, or implement-house, there might be a lock-up workshop, with door to open into the place, with a few tools, paint pots etc. the expense of which would not exceed five shillings. The man should be encouraged to make his duty a pleasure, and to feel a pride in showing his master's implements in fine order. The waggoner might be the most proper man to be the farm mechanic, and he would also have the opportunity of getting what he wanted when at the market town.

Turnip ridge roller

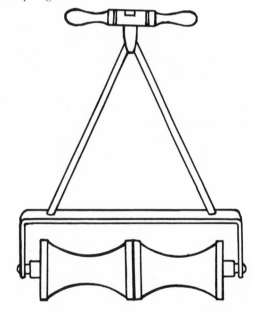

3

Steam cultivation

Steam Ploughing

The idea of using a rope and mechanical power to haul a plough was first explored about 1800 by Mr Richard Lumbert of Gloucestershire. At that time he was using a **mole plough** for drainage purposes and he devised a framework with double drums to wind in the rope, during which time it hauled the plough through the soil. The frame was anchored to the ground and the motive power on the drums was a team of eight women.

Other inventors had explored the possibilities of using a mechanical power long before 1800, the earliest patent for cultivation being taken out by David Ramsey in 1630. His specifications were vague but his invention, 'for making the earth more fertile', included some form of land transport with ploughs or other implements hauled at the rear.

Nearly 140 years elapsed before Francis Moore patented his 'Fire engine to supplant horses'. Again no clear specifications were given, but the inventor held a strong opinion that horses would be quickly superseded by steam power as a means of traction both on farmland and public highway. Richard Lovell Edgeworth devised a steam engine that travelled upon an 'endless railway system'. It was not a railway system with tracks as such, but simply a number of flat bearers attached to the wheels of the engine in order to support its weight on soft land. He patented it in 1770, but the endless railway method was not successful until the **Boydell patent** in 1846. The interest shown by various inventors was enough to make James Watt himself consider it necessary to safeguard his own ideas on steam cultivation in 1780. This covered the propulsion of steam carriages, and the arrangement of agricultural implements behind the carriages.

The earliest patent for a steam haulage system, wherein the plough or other implement was not attached to the rear of the moving vehicle but dragged along on the end of a rope, was patented in 1810 by a Major Pratt, ten years after Lumbert's mole plough. The Major's system made use of a steam engine placed at one end of the field and a heavy four-wheeled anchor cart equipped with a large horizontal pulley wheel stationed at the other end. The engine had a winding drum beneath its boiler and turned an endless rope around the distant anchor pulley, paying it out and winding it back onto the drum alternately, whilst a plough attached to the rope was caused to move back and forth between engine and anchor. This important idea does not seem to have been carried out or even brought to the notice of other inventors at that time, because practical steam tackle designed on the very same lines did not emerge until a further fifty years had elapsed.

John Heathcoat's machinery. Inventions continued to flourish, but not many of them met with practical success; in fact nothing of outstanding value was achieved until John Heathcoat, lace-maker, factory owner and Member of Parliament for Tiverton, Devon, obtained a patent in 1832 for 'certain new improvements in draining and cultivating land, and new improved machinery which may be applied to divers other useful purposes'. He intended his apparatus for bog reclamation and ploughing work on marshy land, a situation too difficult for any horse-drawn plough, and consequently designed his winding engine with a platform of endless track on both sides, each track turning around large drums at the fore and rear end of the engine, with intermediate tensioning pulleys placed between them. The distance between the drums was 26 ft. and each track was 6 ft. in width, so that the resulting supporting area of about 300 sq. ft. supplied the necessary bouyancy for the huge machine to work on waterlogged land. The system required the engine to be placed about 300 yards away from, and in line with, the anchor pulley and a plough to be attached to the drag rope between the two items. The rope was wound in and paid out alternately so the plough moved to and fro continually between engine and anchor, both of which were moved across the land as ploughing progressed. John Heathcoat later extended the system to include two ploughs instead of a single one, and to achieve this, two separate anchors had to be provided, one on each side of the engine. The engine was gradually moved across the centre of the land and the two portable anchors across the headlands in the same direction, with the two ploughs working continuously between the engine and respective anchors, until the land was completely ploughed over. He used a specially designed **balance plough** to cut double furrows, a style of plough which continued with cable cultivation for the whole of its duration, except that as the power of engines was increased so the plough was made to cut four furrows instead of two. Such ploughs required a person to be seated upon the beam to steer the wheels and correct the

Heathcote's steam plough, 1836

deviation, whilst the remainder of Heathcoat's system required another ten men whose attention was divided between the engine and the rest of the apparatus. The inventor was only one of many who lost a great deal of money (a reported £12,000) on such an enterprise and as efficient as his machine and method seem to have been, the reclamation of marsh land did not capture the imagination of agriculturalists or rouse much attention at all except for the approving comments of men like drainage expert Josiah Parkes, who assisted Heathcoat with much of his work. Had the inventor concentrated upon, or at some point transferred his attention to, the cultivation of normal land, he no doubt would have achieved excellent results and also gained many of the rewards that went to John Fowler during the second half of the nineteenth century.

Some other pioneers associated with early steam ploughing before John Fowler entered the scene, were Messrs Fiskin of Newcastle, John T. Osborn and Lord Willoughby d'Eresby, who presented steam tackle at the 1851 Great Exhibition. Lord Willoughby's was a double engine set which incorporated two winding engines placed opposite to each other some distance apart. The drag rope was attached to both engines and it was first wound in by one and then by the other so that a plough fastened midway along the rope was moved to and fro between the two engines. Osborn's system also used double engines and each one contained two winding drums. The engines were positioned opposite to each other and the ropes were paid off and wound onto the other engine's drum respectively. Each rope had a plough attached to it and both were at work simultaneously, passing at a point midway between the engines. This inventor was not the first to propose that iron rails should be laid down each side of the field for the two engines to travel upon. This idea was brought up repeatedly by different people, but whereas permanent tracks were quite suitable for the railway systems then being developed, they were hardly flexible enough for farming, and once installed about the land would have been more a hindrance than a help. The **Boydell railway** invented in 1846 finally relieved the concern attached to moving heavy engines about on soft or marshy ground, and about the same time iron-wire rope effectively replaced link chains and the ordinary ropes used for plough haulage. Direct traction or rotary digging machines, in which the engine that supplied the motive power accompanied the ploughing implement across the field, began to emerge alongside the steam haulage systems about this time. Some of them were highly

91

successful and were put into the work of farming for a short period, but many people, especially inventors and engineers, were very concerned about the loss of power with this method, whilst the majority of farmers were convinced that it offered no advantage over the common method of horse ploughing. However more direct traction machines were introduced and opinions became very much divided between direct traction and cable haulage systems. There were advocates for each, and especially for the latter was a gentleman named Mr Hannam of Berkshire whose cable system was first manufactured by Messrs Barrett and Exall of Reading in 1850. His was the 'Roundabout' system, later used so very effectively by John Fowler, in which the winding engine stood at one corner of the field whilst the plough traversed across the field between the two portable anchors, which were gradually moved down the headland towards the engine, the cable gradually being reclaimed onto the winding drum as work proceeded. Only sixty acres of land had been completed before Mr Hannam's engine and apparatus were forced to halt because of repeatedly broken cables. This was an important consideration since a 500 yard length of cable, woven entirely or iron wire, cost in the region of £50. Later a single strand of steel wire was interwoven with the iron wire, but unfortunately for Mr Hannam, all-steel cables, which were very much stronger than the iron ones, were not introduced until 1857.

John Fowler's apparatus. The first inventor to demonstrate that cable cultivation could be practically and profitably employed was John Fowler, who in 1852 specified a stationary steam engine to wind in a rope and draw a mole plough with it. It worked successfully and established a new trend in methods of land drainage for which he was awarded a prize by the Royal Agricultural Society at their Lincoln Show in 1854. He was encouraged to apply the same principle to cultivation of the top soil and, no doubt taking a lead from Mr Hannam, he worked on the idea and within two years his apparatus, manufactured by Ransomes and Sims of Ipswich and Stephensons & Co of Newcastle, gained the Gold Medal of the French Agricultural Exhibitions. It was duly awarded the long-offered Royal Agricultural Society's prize of £500 at the Chester Show of 1858, and by 1860 the inventor and his apparatus had won even wider acclaim and a total of £1,065 in prize money. In that same year, at the occasion of

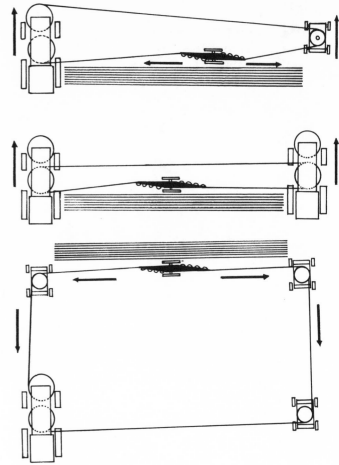

1 Single-engine system
2 Double-engine system
3 Roundabout system
Diagram of Fowler's ploughing system

the Canterbury 'Royal', his apparatus was subjected to the most rigorous tests which undoubtedly established its practical value over competitors before the public.

Fowler's steam ploughing apparatus was made according to three distinct systems. In the first system there was a single steam-driven winding engine stationed at one side of the field, whilst on the opposite side of the field there was an anchor which formed the point of resistance for an endless steel cable, alternately played out and wound in by the engine. The anchor was a heavy wooden platform containing a large horizontal pulley wheel around which the steel cable was turned, and the platform rested upon four sharp-edged iron discs which sank into the soil and prevented sideways movement of the anchor under the strain of ploughing. Until Fowler developed his **self-adjusting anchor,** the former less efficient anchor had to be moved forward manually, and both it and the winding engine progressed slowly along their respective headlands, always keeping exactly opposite and at the same distance apart, whilst the plough attached to the cable traversed to and fro across the field between them. When not engaged in the work of ploughing, the engine could propel itself from one field to another or along its common highway, but it required a single horse in shafts for steering. The cost of Fowler's steam ploughing apparatus was quite beyond the means and also the requirements of the ordinary farmer, as the following price quotation taken from an 1860 Ransome's catalogue will indicate. The tackle consisted of one double cylinder 10 h.p. winding engine with self-moving and reversing gear, one water cart, one portable anchor, 800 yards of steel cable, cable porters, snatch blocks and field tools etc. The cost of the equipment was £622, with an additional £81 expense required for the four-furrow balance plough. A **cultivator, subsoil plough** and **harrows** especially designed for Fowler's set was also available for purchase. The subsoil plough and cultivator were particularly recommended to break up 'plough pan' followed by the layer of soil just beneath, the normal depth of plough penetration which had been gradually but firmly compressed by the weight of plough and draught-animals over countless generations. The effect was to break up soil for the benefit of plant roots and the importance of land drainage. The mole-plough was even more effective for the latter purpose when drawn by steam power and its use became essential for the successful draining of clay land where the cost of laying tiles or pipes was prohibitive. Despite the high cost of steam ploughing apparatus, thirty-five complete sets of Fowler's tackle had been ordered by 1860, many for farmers in England and Scotland who intended to hire them out for work on neighbouring estates. Only three years later there were estimated to be some 600 sets of steam tackle in regular use up and down the country by contractors, many of whom were not able to survive the following decade of successive rain and falling prices. Few of them managed to succeed in the arable areas of England and Scotland but, in general, steam power was most commonly used for threshing and barn work until later in the century.

Fowler's second system came as a result of his experiments with the design of winding engines and layout of the anchors etc. It was perfected about 1860 and was destined to become the most effective and best-known form of steam cultivation in the world. It required the use of two portable winding engines placed at opposite ends of the field, and they acted alternately: one played out the cable whilst the other wound it in, the ploughing implement attached to the cable being dragged from one end of the field to the other alternately whilst both engines gradually moved along their respective headlands. This system was even more expensive to purchase and maintain than the first, but it was recommended for two reasons. Firstly, the speed and ease with which the tackle could be set to work as soon as the engines had reached their destination, and secondly, the engines were capable of transporting themselves and the whole of the tackle to the place of work, requiring no additional manual or animal labour.

The third system was a 'Roundabout' in which the winding engine was stationed in one corner of the field and the anchors at each other corner. The plough traversed the field between two of the anchors, which were gradually moved down the headlands towards the engine.

Fowler's ingenuity was rewarded, and by the early 1860s his factory employed over 1,000 men, all working on the production of various argicultural implements. He died at the age of thirty-eight, but his ploughing systems continued to provide the basic pattern for cable haulage until they were totally replaced by tractor-drawn ploughs in this present century.

The advantages of steam power as a means of tillage were slowly being recognised, but the cost of all the necessary tackle was far too high for the majority of farmers. However it was not long before steam ploughing companies began to form in different parts of the country and they were able to complete the work on a contract basis at about half the cost normally required for horse ploughing per acre. Those farmers who owned such apparatus were fortunate in having a considerable power to hand at the moment their land was in a suitable state for cultivation, and used at the appropriate time even the most obstinate land could be thoroughly broken down for the reception of seeds.

Some other interesting systems of cable cultivation were devised just after the middle of the nineteenth century, and whilst the majority did not pass the experimental stages others were practical enough to work in competition against Fowler's machines.

The Marquis of Tweeddale was responsible for steam tackle first put into operation in 1857. It was manufactured by Tullock's and employed double engines in the manner of Fowler's second system. They were however designed in quite a different manner and instead of having the winding drum placed beneath the boiler, the Marquis had it mounted in a vertical position at the rear of the engine chassis. The fore carriage wheels could be turned to effect steerage and travelling motion was supplied by the engine driveshaft to the rear axle. Two ploughing speeds were obtainable, $3\frac{1}{2}$ and 5 m.p.h. A huge double furrow turnover plough was drawn back and forth between the two engines, with two shares working the soil whilst the other two were in the air. When the plough approached an engine at the end of its run across the field, it was lifted clear of the soil by means of a cranked shaft attached to a horizontal beam above the winding drum. Whilst the plough was raised up it was turned over on its axis to bring the opposite set of shares into work, and was then lowered for return across the field by a further half turn of the crankshaft. This way the plough was turned about, almost as fast as if it were drawn by horses. When the plough was moving away from either engine, that engine moved 56 in. forward along its headland or equal to four 14 in. furrows, and so the engine assumed the correct position for the plough's return. When this tackle was moving between one field and another, the overhead crankshaft was taken down and together with the plough and cable was loaded on to a wagon at the rear of the engine, the whole procession having a roadspeed of 4 m.p.h. The inventor was perhaps a little too cautious in believing it necessary first to go over the land with a horse-drawn **subsoil plough** in order to break it down and then to clear away stones by hand. However subsoiling would have greatly improved the land drainage.

Ploughing, cultivating and harrowing were increasingly completed by steam as the century progressed but the belief that steam cultivation would supersede horse work had not been realised. Many farmers still thought that the large implements employed in steam cultivation were too heavy and too clumsy to work properly, besides treating the land too roughly. Where steam cultivation was employed it was generally done like steam threshing, by contractors who owned the tackle and travelled from one farm to another doing the work for an agreed price. There were many who said that steam cultivation was designed for heavy soil only, and that it was not necessary on lighter land anyway. Special implements however were designed in order that farmers with light soil might take advantage of steam cultivation, if they so wished. Many new manufacturers entered the field and each added something of importance to the stream of inventive ideas.

Messrs Coleman and Morton gained the Royal Agricultural Society's Gold Medal in 1864 for their new ploughing engine. It had two small drums mounted vertically beneath the boiler, and each rotated in an opposite direction to the other so that as one was paying the cable out, the other was winding in. This was a single engine system with the cable passing around a distant anchor pulley.

John Allin Williams of Baydon, Wiltshire, in 1855 patented a machine fitted with a pair of windlasses which were driven by steam power at a slow rate from an ordinary **agricultural locomotive.** The windlasses were contained inside a framework upon travelling wheels and could be attached to the steam engine by a belt whenever required. This relieved the farmer of the need to purchase a special winding engine.

J. & F. Howard of Bedford became increasingly involved in the manufacture of winding engines and steam tackle, both of their own design and that commissioned by special order. One of their earliest successes was a combined drain cutting and sub-soiling outfit that made use of the separate windlass attached to an ordinary steam engine, both of which stood in one corner of the field with the cable passing round four separate anchor pulleys. It worked successfully and more than forty sets of this tackle had been sold by 1860. They continued to patent other systems and by 1861 had developed a new system of cultivation whereby two twin engines were placed on opposite headlands and each worked one half of the field.

Thomas Aveling produced his first chain-driven ploughing engine in 1870, and two other famous manufacturers, Tasker's of Andover and McLaren's also took up manufacturing about the same time.

The Savage Agriculturist. In 1880 Frederick Savage of King's Lynn, Norfolk, perfected an ingenious method of coiling the cable inside the wheels of an engine instead of onto separate winding drums. A similar sort of thing had been patented

Two views of the Savage ploughing engine. From the
manufacturer's catalogue

Single-engine system of ploughing with a Savage engine

by Messrs Chandler and Oliver of Hadfield in 1856, but Savage improved upon it and then incorporated the idea into his single cylinder ploughing engine, known as the Agriculturist. A side view of the engine is illustrated by an engraving from Savage's catalogue, and there is also a view of the same engine, showing the winding drums incorporated in the large wheels, the rear end of the engine being raised clear of the ground by packing. The system worked by means of the cable being coiled onto the wheels in opposite directions: one was caused to pay out whilst the other was winding in. To reverse the system, the driver simply had to rotate the wheels in the opposite direction. When running on the road as an ordinary traction engine the cable was removed and the winding grooves in the rear wheels covered by the quickly detachable plates, which can be seen piled beneath the engine boiler. The pictorial woodcut gives an aerial view of the Savage Agriculturist operating a single engine system in the field. A large number of these engines were made at the St Nicholas Works in King's Lynn, and many were exported to the Continent.

The vertical or side drum engine became more prominent at about this time. Although the idea of mounting the winding drum in a vertical position at the side of the boiler barrel had been considered much earlier by an engine maker named Weeks, it was not until 1879 that it was incorporated into a patent by Everet and Adams. Their ploughing engine was made by Chas. Burrell and Co at Thetford, and utilised either a single drum or two drums placed one either side of the boiler barrel. This type of ploughing engine was known as the Universal and was used for both double and single engine systems. Quite a number of manufacturers brought out a Universal engine including Fowler's and Garrett's. Many of them were shipped to the Continent.

Some steam tackle was also shipped to America and both single and double engine systems were employed for a short time. They were found to be less effective on the dry soil there, and the idea was quickly abandoned. In 1894, R. L. Ardrey, the American agriculturalist, observed steam ploughing to be clumsy and added that 'the more popular plan has been to draw a gang of plows behind a traction engine. In some cases a modified form of threshing engine has been used of sixteen or twenty horse power or larger. Excellent results have been obtained

and many outfits of this type are now in use.'

The increasing demand for ploughing and harvesting engines in America brought forth the development of a tricycle form of engine, wherein the weight of the boiler and mechanism were carried upon two large, wide wheels, with a third wheel at the front for steering. They were capable of developing 40 to 80 h.p. and could draw eighteen 12 in. ploughs to complete about 50 acres in one day. By 1895, these three-wheeled engines were in general use on large wheat farms in the West of America, but this form of engine was not popular in England owing to the difficulties encountered with the overloaded wheel at the front end which tended to sink too deeply into soft soil. These engines were mostly built by Jacob Price at the J.I. Case works in Racine.

For some years following the introduction of oil tractors, European manufacturers promoted a cable system, centred upon their machines. It did not follow directly in the tradition of steam cultivation, as the cable was anchored at both ends of the field and the motor plough or tractor with plough attached at the rear wound itself back and forth from one anchor to another alternately. Both anchors were moved along their respective headlands until the whole of the field was completed.

Some early oil tractors were equipped with a cable winch and about 300 yards of cable. The tractor released the cable as it traversed the field to the opposite headland and upon reaching that position the machine was anchored and then proceeded to wind in the cable drawing a plough towards itself. If the length of the field exceeded that of the cable and many hedgerows had been removed to make the fields larger for steam ploughing, the tractor would halt when the whole of the cable had been played out, and complete the lines of ploughing in convenient stages. Tractors equipped with a single windlass were well suited for trenching, subsoiling and mole ploughing. The implement was hauled towards the tractor only, during which time it worked the soil and upon completing the line was returned to its former position by horses, in readiness for another run.

Two single windlass oil tractors were occasionally used as a double engine system, drawing the plough to and fro between them, but they were not as efficient as steam engines because their distance apart was limited by the smaller horse-power of the motors, and consequently the shorter length of cable. During the 1920s in Britain and France, some experiments were undertaken to determine the suitability of electric power for cable cultivation, but due to the ever-increasing efficiency and versatility of direct traction methods, together with low maintenance and operative cost of the tractor, it was pushed into the background. By 1925 cable cultivation had completely given way to direct traction methods, where the implements were attached to the rear of a tractor. After that time steam power was used. only in such difficult situations as defeated the tractor and for special operations, such as subsoiling, trenching, and mole ploughing. Some farmers, who retained their tackle, unwisely, converted the steam engines to diesel power as the price of coal gradually became more prohibitive. Other tackle was procured by collectors and by enthusiastic young farmers who have restored it to near-perfect condition.

The Instruments used with Steam Cultivation

The balance plough could turn from two to eight furrows simultaneously, according to the power of the engine and nature of the land. Early versions were constructed of timber except for the iron digging parts, but it became increasingly necessary to provide them with a rigid iron framework as the efficiency and power of the winding engine was increased. This also applied to the other implements designed for steam cultivation. The illustrated plough, by J. Fowler and Co., about 1870, with its angular beam balanced at the centre upon an axle and two wheels, is a typical arrangement. It had two sets of ploughs attached to the beam, four at one end pointing towards those at the other, and there was also a driver's seat at either end of the beam, where a man steered by altering the angle of the travelling wheels through a worm and rack mechanism. It was usually arranged so that extra plough bodies could be bolted on when required, their position being adjusted by set screws or wedges whilst in addition subsoiling tines could be attached behind each plough for the purpose of breaking up the lower strata to a regular depth without turning it into the top soil. Upon reaching one end of the field the implement was prepared for its return journey simply by pulling down the end of the beam and ploughs that had been suspended in the air and directing them back into the next line of work.

Harrows and cultivators were built on a large scale for use with cable cultivation, and were often 10 ft. or more in breadth so as to render them capable of covering many acres during a day's work.

Steam balance plough

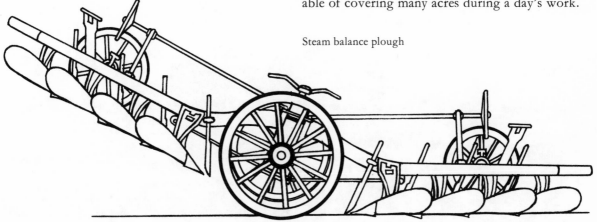

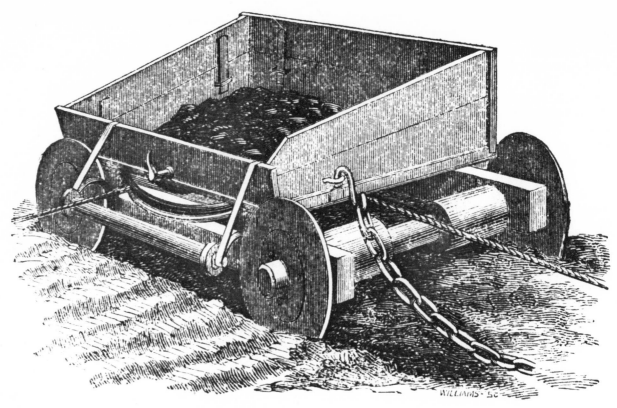

Fowler's self-adjusting anchor. From Ransome's catalogue

Burrell-Boydell traction engine being inspected by the Tetney
and Louth Agricultural Society in 1857

Disc harrows were increasingly used in steam cultivation during the last quarter of the nineteenth century. They were constructed in similar manner and dimension to the horse-drawn version, or alternatively the rows of discs were slung on chains beneath an iron framework on four wheels.

Iron land rollers were moved in a similar manner and with the power of steam available they could be made 15 ft. or more in breadth.

A **multi-purpose framework,** an iron bridge-like structure supported at the fore and rear ends by wheels, was produced by some manufacturers. Two or three disc harrows or the same number of rollers of any pattern could be attached beneath the framework by means of chains fastened to their axles, with their whole weight bearing down upon the ground. The driver was perched up above and could steer the wheels at either end as the arrangement was hauled back and forth across the field.

Corn drilling was successfully completed by the owners and operators of steam tackle. A popular drilling arrangement was a heavy harrow, a seed drill and a light harrow placed in that order beneath a bridge framework. The heavy harrow produced a firm tilth, the drill dropped the grains into the soil and the light trailing harrow covered them over.

All of these implements and other oddities designed for cable cultivation were abandoned as oil tractors gradually became established.

Fowler's self-adjusting anchor is illustrated. The cable from the steam winding engine on the opposite headland passed around the horizontal pulley wheel, and as the plough was drawn back and forth between engine and anchor the motion of this pulley was transmitted through gearing to the carriage wheels, causing the anchor to proceed along its headland in constant line with the engine. Only the minimum of supervision was required, leaving men free for other work. The cart-like body of the anchor was left open to receive stones and rubble during land clearance or for additional weight. The thin iron wheels were designed to sink well below the surface of the soil and prevent sideways movement.

Boydell's endless railway or patent wheel was attached to many steam traction engines after 1846. It was devised in order for the machines to cope with wet farmland, a condition which normally gave rise to wheel spin and loss of power. The railway did not include tracks but consisted of six beams or bearers attached to each of the engine's rear wheels.

The illustration shows the position of the bearers around the perimeter of the wheel, where they were loosely retained by V-shaped retaining rods and cross rivets. As the wheels revolved, the bearers fell singularly upon the ground and formed a support or continuous platform for the vehicle. Tuxford, Bach and other English manufacturers incorporated this feature when building steam engines for Chas Burrell and Co.

The dreadnaught roadwheel was an Australian development that enabled their steam engines to exert full power on swamp lands. It was invented by an Australian, Frank Bottrill, in 1906, and consisted of wooden beams attached by cables, singularly or in pairs, around the drive wheels of a traction engine. As the wheels revolved, each bearer came into contact with the ground, providing buoyancy and support for the vehicle.

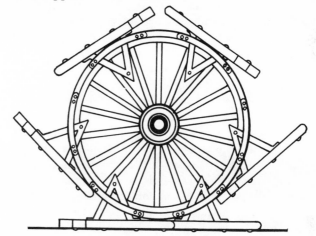

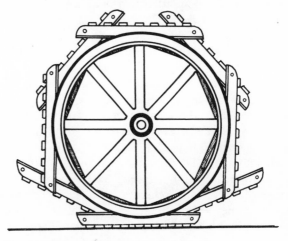

Top: Boydell's endless railway
Below: Bottrill's road wheel

99

Steam Digging

During the nineteenth century there were many attempts to perfect rotary digging machinery but most were largely impractical owing to the fact that most inventors in this field were in no other way connected with agriculture and therefore did not anticipate the problems involved.

The earliest digging machines were designed for horse-power rather than steam, and perhaps the first to be recorded was the **drain cutting wheel,** mentioned by the Board of Agriculture in its report on land drainage in 1801. Briefly, it was a large iron wheel with sharp digging tines set around its circumference. It was held vertically by a framework and after being lowered into a hole, previously excavated by hand, was made to spin rapidly by the power of horses and capstan. The rotating tines threw the soil back and out of the trench as the wheel advanced, leaving a narrow slit trench behind. This machine was probably never constructed but similar wheels appeared later in the century.

In 1846 Thomas Bonser of Merton, Surrey, and William Pettit of Lambeth combined their efforts to patent 'certain improvements in machines for tilling land'. Theirs was a kind of rotary plough and instead of a mouldboard it had a spiral or screw device with radial cutting blades attached below the frame. As the plough was drawn across the field by horses the screw revolved and cast the earth to one side, leaving a channel in its wake. The inventors did consider steam power, but it would seem they used horses instead. Bonser and Pettit also patented a horse-drawn rotary tiller that made use of an iron cylinder armed around the outside with a spiral arrangement of blades or tines, and was mounted across the rear of the machine at right angles to the line of draught. The cylinder was rotated into the direction of travel by means of gearing from the wheels, so that the spikes entered the earth as they descended, then threw the shattered soil backwards as they emerged. This idea was later improved by other inventors, notably Romaine in 1853. Another early effort was Thomas Vaux's revolving harrows of 1836, followed by Joseph Hall in 1842, but none of these early schemes were ever extended beyond the experimental stage. The revolving harrows were much used for land cultivation in later years, when they were known as **Norwegian harrows** and made suitable for horse traction. The idea of a screw plough later emerged in the Archimedean subsoil plough, and the revolving wheel laid a foundation for successful pipe laying and land drainage machinery. Other early rotary diggers included Paul's drain cutter of 1847, a large revolving wheel armed with tines. A machine by Sir John Scott Lillie was designed so that its tillage portion worked by gearing from the land wheels or by a steam engine mounted upon the carriage, whilst the carriage propelled itself to and fro along a suspended cable. There was also Mr Bethell's 1857 system of revolving forks mounted upon the rear of an agricultural locomotive, itself fitted with **Boydell's endless railway,** a track-laying device particularly invented to avoid heavy engines compressing farm land. The first major digging machine was James Usher's rotary steam plough made in 1849. The inventor lived in Edinburgh and his machine was tried out in southern Scotland, where it performed well, but its purchase price of £300 placed it beyond the reach of all but the richest farmers. It can be observed in the drawing to be similar to a portable steam engine of that time, but it differed in that the boiler and duplex cylinders were mounted upon two rollers instead of wheels. One roller supplied steerage, whilst the other transmitted power from the engine to the rotary digging device at the end of the chassis, where curved mouldboards and shares were placed side by side along a horizontal transverse shaft. The shares rotated and penetrated the soil in the opposite direction to which the machine travelled and with the two shares touching the soil at the same time, even tillage was assured. This 10 n.h.p. engine weighed 5 tons and worked at an average rate of about 2 m.p.h., ploughing up a strip of land between 36 and 50 in. wide to a depth of 9 in. at a cost of 2s. 6d. per acre. The same work completed by horse-drawn plough would have exceeded 8s. per acre.

In 1857, Thomas Rickett, a steam carriage manufacturer of Castle Foundry, Buckingham, entered the rotary tillage field with a 10 n.h.p. machine. He employed wide wheels instead of rollers and again the power was produced in duplex cylinders, this time enclosed at one end of the boiler. The digging device positioned horizontally across the end of the machine was an iron spiral or screw, which was rotated at high speed by chain drive in opposite direction to the traction wheels. It was designed to cultivate soil to a width of 7 ft. 6 in., but it is doubtful whether it ever reached a really practical state.

Rickett's cultivator. From the R.A.S.E. Journal, 1858

The guideway system was a remarkably progressive, but exceedingly costly, invention by Lieutenant Halkett, R.N. This proposed method of cultivation proved highly impractical as it necessitated the installation of permanent railway tracks, placed in parallel across the land at intervals of 50 ft. Two steam engines, placed on adjacent tracks, were connected together by a light girder work bridge, which was suspended well above ground level so that various implements, ploughs, cultivators, drills, harrows and hoes etc., could be attached to its underside where they completed the work. There was also provision for transplanting and watering in the plants. The two engines travelled along the tracks at exactly the same speed, carrying the bridge and implements with them, proceeding back and forth until the whole 50 ft. wide stretch of land was cultivated. Then the whole arrangement was moved over to bridge another section of land by means of link rails. The inventor claimed that working costs

were little more than a few pence per acre, but however ingenious it was, the cost of installation was prohibitive to most.

Romaine diggers. In 1853, John Joseph Mechi of Tiptree Hall, Essex, financed the first of three spectacular Romaine diggers developed by Robert Romaine of Peterborough, Canada. This machine moved on wheels but was drawn by horses; only the digging part, a large iron cylinder armed with curved iron spikes, was powered by steam. A second version of this machine was shown at the Great Exhibition of Paris in 1855, where it caught the interest of William Crosskill, the renowned agricultural implement manufacturer of Beverley, Yorkshire, who agreed to collaborate with Romaine upon an improved and fully steam-powered machine. The resulting Romaine-Crosskill digger of 1857 moved on two wide wheels, whilst steering was effected by means of small castor wheels at the fore-end. The digging cylinder, which remained much

Halkett's guideway system. From *Quarterly Journal of Agriculture*, 1858

Romaine-Crosskill digger

the same as before, a hollow iron cylinder 2 ft. 6 in. in diameter by 6 ft. 6 in. long, covered with curved iron spikes. It was situated across the end of the machine where it dug into the soil for a depth of about 12 in, both it and the wheels being powered by the 14 n.h.p. steam engine. The large travelling wheels could each be thrown out of gear so that the whole machine turned in a small space on the headlands, leaving little land uncultivated. Despite the considerable sum of money invested by Mr Crosskill and the enthusiasm of many interested parties, the considerable weight of this machine was a severe handicap as it was apt to sink into wet ground, even when fitted with wider wheels.

W. Smith's gang plough. Meanwhile, due to the effort and ingenuity of men like John Fowler, the haulage of tillage implements by steam-powered cable was proving more successful than rotary cultivation. Few original ideas came forward for a time except for the illustrated W. Smith's 1860 gang plough, based upon the use of steam-driven continuous tracks and a platform with seven plough-shares fixed at either end. The vertical boiler and duplex cylinders in the centre of the platform provided moving power to the tracks and the balance of the machine could be tipped on the large wheel at the centre, whereupon its entire

weight was applied to one end and the set of shares working the soil.

Broadside diggers. Thomas Churchman of Pleshy, Essex, produced the first of his huge and effective broadside diggers in 1877. Some thirty of them were subsequently manufactured by Eddington's of Chelmsford, J. & H. McLaren of Leeds and Savage's of King's Lynn, each one weighing a grand total of 20 tons and costing in the region of

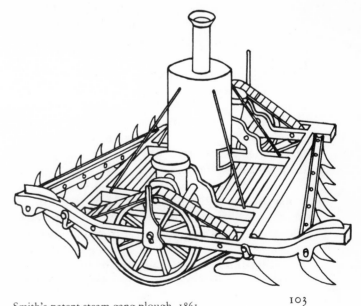

Smith's patent steam gang plough, 1861

103

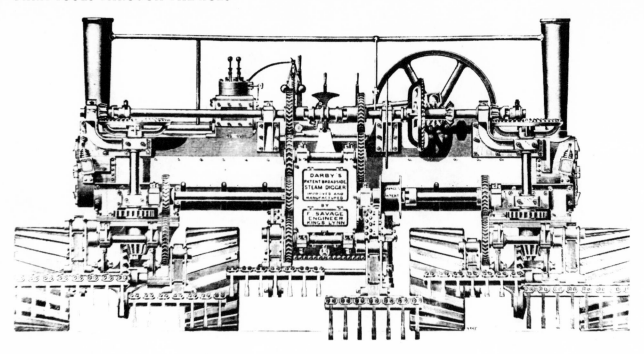

Darby-Savage digger, 1880

£1,200. The Darby digger took the form of a single cylinder engine above a horizontal boiler, with a two-speed drive to the 3 ft. 6 in. diameter travelling wheels, and digger gear which consisted of forty-one tines in three sets, working a width of 20 ft. 7 in. When the machine was digging, the axle of the travelling wheels was parallel to the boiler, as shown in the illustration, and steering was delivered by the hand wheel on the footplate to the eight small wheels projecting at the rear. These machines were too wide to travel along the public highway unless first each end was jacked up in turn, the wheels turned through 90 degrees and the steerage gear dismantled.

The Darby-Savage diggers, later made by Savage's under Mr Darby's personal supervision, can be seen as much the same as the former in general arrangement, though some mechanical refinement based upon past experience was introduced. The new machine had six sets of digging tines instead of three and the working depth was increased from 6 to 9 in. By the use of integral hydraulic jacks, the whole weight of the machine could be taken up off the axles whilst the wheels were turned round and placed in line with the boiler, for travelling through gateways and along the road, a conversion which generally took about 45 minutes to execute. The effect of each tine contained in the frame was to

carve out a furrow 9 in. deep and 6 to 9 in. wide, a total width of 21 ft. being ploughed in one operation. When at work the reciprocating forks entered the ground vertically, though they could be adjusted to suit the condition of the soil.

The boiler of this magnificent machine was 16 ft. 6 in. in length and consumed over 400 gallons of water every hour; the firebox and footplate controls were at the centre and a single cylinder developed 8 n.h.p. to move the digger along about one mile every hour.

Towards the end of the nineteenth century a number of more manageable diggers were developed by some already established engine makers, but they were not greatly appreciated by agriculturalists. They took the form of reciprocating or rotary tines contained in a frame at the rear of a general purpose type of traction engine and therefore were not as awkward in use as the broadside diggers, but again the combined weight was a severe handicap.

By that time, steam cable cultivation was very much in vogue and the new diggers, as interesting as they were, did not catch on but were relegated from the agricultural scene to the task of highway construction and similar work where further development was rapidly encouraged and subsequently brought about in this form of machinery.

The idea of rotary digging was not altogether abandoned in agriculture, although manufacturers' enthusiasm had waned with McLaren's remark that the common plough would always be more efficient than any rotary tillage machine. It was not until the internal combustion engine became available that much further progress was made. Fowler's company produced a 23 ton, 255 h.p. gyro-tiller in 1927. It was a caterpillar tractor at the rear of which were two rings of shares, which rotated upon a vertical axis and ploughed up soil to a depth of 20 in. Similar ideas were explored by inventors in different countries, and various sizes of machine were trialed. But as yet rotary machines are used mainly in market gardens and for small acreages.

The Rotary Hoe

Arthur Clifford Howard, an Australian apprentice engineer, witnessed the occasion in 1910 when a steam tractor was first used on his father's farm at Gilgandra. The tractor fascinated him. It worked well, but he observed that it lost a great deal of power before the wheels could gain enough grip to draw a plough. He imagined that if the power of the engine could be directed to a spinning tillage system instead of to the wheels, this wasted power could be used to the farmer's advantage. For two years the idea lingered, then in 1912 he used a test drive from the tractor to the shaft of a one-way disc cultivator. The rims of the discs were notched so that they ripped into the soil, the transmission was a concoction of chains and cogs taken from machinery around the farm. He immediately achieved some success in cutting down weeds and moving soil but both, along with small stones, were scattered in a very dangerous fashion for some distance. He then applied his efforts to evolving a better shape of blade that would till the soil effectively and reduce throwing. After experimenting with blades of different shape and dimension, with the help of fellow apprentice Everard McCleary, he produced the now familiar L-shaped blade. Using this pattern of blade he built a small rotary hoe machine powered with his own motor-cycle engine, only to find that no market existed in Australia for a tillage implement on such a small scale. At that time, Chinese labour was cheap and the big wheat farmers showed some interest, but required much larger machines

The original rotary hoe, 1920

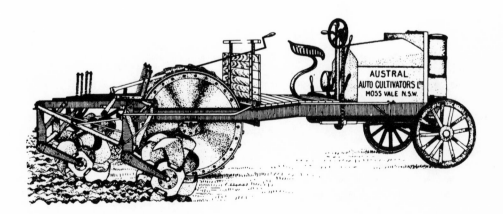

which would equal the work of their big horse teams. Howard and McCleary remained undaunted. They were on the point of producing large combined cultivating and seeding machines when World War I broke out, carrying them both to England, McCleary to his death, and Howard to the Napier's experimental department. When the war ended Howard returned home to continue his venture with the support of a few friends and their finance. He then designed and built a machine especially for the wheat farmers which was completed in 1922. It had a 60 h.p. Buda engine especially imported from America and was built in five sections with a total width of 15 ft. It had three rotor speeds and six travel speeds, and on test it cultivated at the rate of $3\frac{1}{2}$ acres an hour. For some years he was obliged to travel about, working and demonstrating his machine. Six Buda-engined hoes were made and sold in 1923, but capital was so limited that his effort turned to making hoes suitable for attachment to other makes of tractor. At that time Fordson was the best available. Its gears were unsuitable, but correspondence between A. C. Howard and Henry Ford resulted in a set of gears being adapted to Howard's requirements. From then on he achieved complete success, devoting his experience to producing hoes for various kinds of tractor, hoes for orchard and vineyard work, for land reclamation and growing sugar cane. Hand-controlled models, including engines, and the Howard D.H.22 tractor were developed. This was the start of large scale tractor production in Australia, and soon orders began to flood in. The Rotavator Company expanded, subsidiary companies were formed in America and now nearly every country in the world uses their implements.

4

Sowing and planting

The Seed Drill

The broadcast sower carried his seeds in a basket, suspended by straps around his neck, and scattered them freely as he walked along. His action was smooth and unhurried, in order to deposit a regular amount of seeds over the surface of the land and to avoid patchiness in the resulting crop. In this case there was no practical way of dealing with the weeds that grew up amidst the plants and often the crop was smothered or undernourished. Nevertheless this traditional method of sowing was practised until the second half of the nineteenth century. Dibbling the seeds became popular during the eighteenth and nineteenth centuries and lingered on after that time in some eastern districts of England. Seeds were planted a few inches apart in straight rows with the aid of a **hand dibber,** and the resulting spaces between the rows of young plants facilitated the use of a hand hoe to kill weeds and open the soils for admission of air and moisture. There was also the **drill roller,** so much used at one time for sowing the fields in Norfolk. Its function was to leave parallel channels, some 4 in. apart across the land, into which seeds were cast by hand and then covered by the passage of a fine toothed harrow. Some farmers used a plough to make shallow drills for the reception of seeds and they were loath to discontinue this even when a good variety of seed drills became available.

Drill husbandry, in which seeds are planted in equidistant rows on flat or ridged ground, between ridges or in the bottom of furrows, was introduced in England about 1730 by Jethro Tull, whose ideas and practice caused much controversy amongst agricultural writers for many years afterwards. Other seed dropping machines had been made before this but their usefulness cannot be gauged with accuracy, so credit must go to Tull for the first practical seed drill.

The ancient husbandmen of China, India and Japan almost certainly dibbled or drilled their seed by some means or another. The ancient Chinese drill is believed to have been a wheelbarrow arrangement which left three shallow furrows in the soil as it went along. Grain descended from the hopper and was dropped into the furrows through spouts. **The Babylonian plough** had a seed dropper attached to its side which was simply a vertical tube down which grains were dropped one at a time

Dibbling

into the channel opened by the plough share. This simple arrangement must have resulted in an enormous saving of grain, but its use does not seem to have been at all extensive. The ancient Sumerians had a seed dropper attached to their plough beam so as to deposit seed in a newly opened furrow as the plough moved along. A similar device did not emerge in European agriculture until the late eighteenth century.

In 1566 a seed drill was invented by a Venetian, Camillo Torello, but his machine remains a mystery. Before the end of that century, Tadeo Cavalini of Bologna had produced a seemingly efficient seed drilling machine. It was a hopper filled with seeds and drawn along on two wheels, and perhaps it was the vibration of the wheels as they went along which caused the seeds to fall through perforations in the base of the hopper and into pipes, down which they were delivered into furrows made by a form of knife coulter. The seeds were then covered by another implement, probably a harrow attached to the rear of the drill. A number of seed drills were

Babylonian seed drill, 1316 B.C.

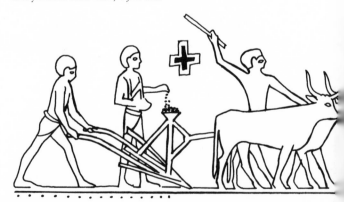

108

patented in England after 1623 and some were described in books and literature of the time, but it is doubtful whether any of them ever reached practical application, including the drill acclaimed by John Worlidge in his 1669 *Systema Agriculturae* as being capable of spreading dry manure, such as pigeons' dung, simultaneously with the seeds. The general arrangement of his drill is illustrated though only a single hopper is present, a second was probably attached whenever necessary to contain the manure. But it was Worlidge's seed dropping mechanism that proved an important contribution to future development. It consisted of a wooden wheel with a number of leather flaps or pockets around the rim. It was made to revolve in the bottom of a seed hopper by means of a belt and pulley from the rear wheel axle, during which the flaps caught up seeds and delivered them as the wheel turned over into the mouth of a curved discharge funnel. This was a primitive kind of force-feed drill, which was later made with a fluted or spirally grooved roller to supply seeds in a regular stream.

An alternative mechanism designed shortly afterwards by a German named Locatelli was the precursor of the spoon or cup-feed drill which became universal throughout Europe. The seed dropper in Locatelli's drill was contained in a separate compartment at the bottom of a hopper and was formed by small metal spoons fixed in four rows along the length of a cylinder or axle. As the axle revolved the projecting spoons caught up the seeds and deposited them into funnels, down which they dropped to the soil. Thus the essential drill mechanisms, the force-feed and the spoon-feed were established quite early, but it is doubtful whether they were ever

Worlidge's drill from his *Systema Agriculturae*, 1669

put into practical use at the time or even known by more than a handful of practising farmers.

Early in the eighteenth century, Jethro Tull constructed his drill to deposit straight rows of seed in the soil at regular distances apart, and his **horse-drawn hoe** with tines set at required intervals to clear away the weeds and aerate the soil between the rows. The son of a Berkshire landowner, he moved to Shalbourne on the Berkshire-Wiltshire border in 1709. There he researched into plant life and the environment best suited to the growth of crops. Despite his extremely poor health and the criticism of all those about him, he persevered and, as stated by Lord Ernle in his *English Farming Past and Present*, 'by constant observation and experiment he learned the difference between good and bad seed, as well as the advantages in care of selection, of cleaning, steeping and change, he also proved that a thin sowing produced a thicker crop and discovered the exact depth at which the seed throve best'. Tull had a small amount of seed sown in channels and

Jethro Tull's seed drill, *c.*1730

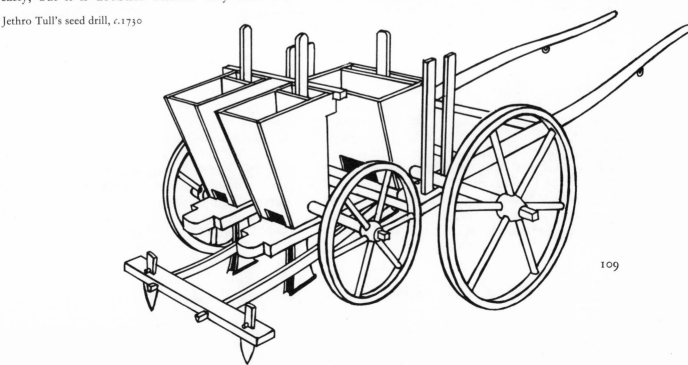

109

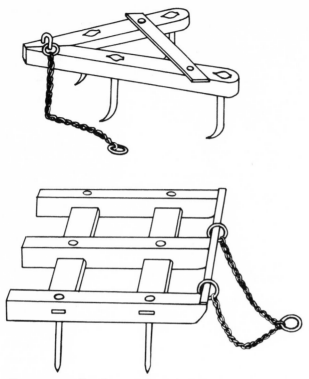

Harrows to trail at the rear of a seed drill

covered evenly with soil. It was successful, so then he set to work 'to contrive an engine to plant sanfoin more satisfactorily than hands could do'. His resulting machine and his system of planting in straight rows was promoted by his book *Horse Hoeing Husbandry*, 1733. It gradually found increasing favour over hand broadcasting and dibbling, but the change was not completed until long after Tull's death.

He designed a number of seed drills, but retained in each the same pattern of seed dropping mechanism, a revolving cylinder with notches or rows of cavities around the perimeter, each one just large enough to receive seeds from a hopper and drop them regularly into a delivery funnel. Tull's three-row corn drill is illustrated. It was light enough to be drawn by a single horse and was carried upon two large wheels at the fore-end and two smaller wheels at the rear. The large wheel axle carried a hopper and turned the centre seed dropping mechanism, whilst the two rear wheel axles carried separate hoppers with their own seed dropping mechanism. The amount of seed passing down to the soil could be regulated by means of a thin brass plate adjusted and held by a spring against the side of the notched delivery cylinder. It worked in much

the same way as the tongue in an organ mechanism, an instrument with which Tull was familiar. The three coulters were arranged so as to pierce the soil beneath the drill and each had a deep groove hollowed into its rear side, for the purpose of directing the seed from the funnels above to the soil below. In his turnip drill he arranged for the seed to be sown at two different depths, hoping that a delayed crop could escape the ravages of turnip fly. What Tull really left for his successors, in the words of Lord Ernle, 'were clean farming, economy of seeding, drilling and the maxim that the more the irons are amongst the roots the better for the crop.' It was not until Tull's principles were put into practice in various parts of the country that their full advantages became apparent.

Other drills were devised after his death, but little further progress was made until 1782, when the Rev James Cooke of Heaton Norris in Lancashire patented his drill that used indented spoons or cups to deliver the seeds. He was in rivalry with another inventor, William Amos, who produced a similar sort of machine, but Cooke's was the one upon which all later drills of the spoon-feed type were based. The earliest improvers of Cooke's drill were Henry Baldwin, a Norfolk farmer, and his bailiff, Samuel Wells, who provided axle adjustment whereby the wheels could be drawn further apart for the addition of more seed droppers, funnels and delivery coulters. They also included a self-regulating lever, which allowed each coulter to ride independently and adjust itself to the irregularities of the soil surface. These improved 'Norfolk' seed drills were made and dispatched, but not in great numbers, to contractors rather than farmers, the majority of whom still preferred to sow broadcast or dibble their seed with the aid of cheap manual labour. Also the labourers themselves were not anxious to be displaced by machinery of any form, which helped to prolong the use of traditional methods.

The practice of drilling was clearly in the interest of better farming, but the spread was by no means rapid. Only the most courageous landowners were willing to adopt it against the wishes of their working men, so advancement was confined to a few areas of England only. General acceptance of the seed drill first began in Suffolk during the early years of the nineteenth century, where only a few years earlier dibbling and broadcasting across the furrows left by a Norfolk **drill roller** had been steadily increasing. Seed drilling was at first only practised by a few, but by the end of the first decade, it had spread throughout almost the entire county of Suffolk and other districts slowly began to follow suit. Further development took place there shortly afterwards when James Smyth, a wheelwright of Peasenhall, Suffolk, commenced business. Together with his brother he devised a method of making seed drill coulters individually adjustable so that they could be set at various intervals apart, in accordance with the type of seed being sown. They also provided swing steerage, a method by which the driver, who walked at the rear of the moving drill, could hold the whole transverse line of seed coulters steady and thus preserve a straight line of planting should the horses or wheels of the drill carriage stray out of line to one side or the other. This principle of swing steerage was later used in the combined seed drill and horse-hoe, wherein the seed hoppers and dropping mechanism were conveniently removed and the same frame and wheels transformed into a horse-hoe. Smyth put a combined 'Suffolk' corn and manure drill into the market. It quickly became popular, but he was not content to provide for farmers in his locality only and, after demonstrating over a wide area, he undertook drilling on a contract basis for a great many miles around. This proved a highly successful venture and his firm expanded. The effects of his work were felt all over the country, but soon other manufacturers came into this field and established themselves with equal reputations.

The drop drill, provided with a mechanism to distribute seeds and manure at desired intervals rather than in a continuous stream, appeared in 1839. Subsequent drop drills could be quickly converted to sow at intervals, or continuously, whichever was desired. The steerage seed drill emerged about the same time. It had an ordinary drill mechanism but was characterised by the addition of a fore-carriage with two small wheels, which enabled the driver to keep in line with the wheel tracks made during the last round of the drill and thus ensure the rows were perfectly straight and parallel over the entire surface of the field. At the Royal Agricultural Society's Bristol meeting in 1842, it was noted that:

Drills have already penetrated into districts where three years since, their purpose and even the name of the implement were scarcely known. It would indeed appear that unless some new

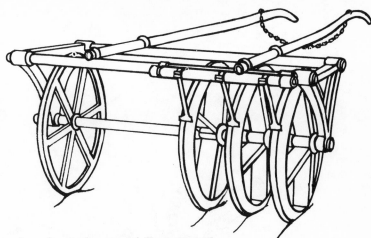

Late nineteenth-century drill pressing roller

principle be struck out which shall greatly excel the present system of construction, little remains to be done as regards fulfilling the required objects of this most important class of implement. Seed corn of every description may be deposited with the greatest of precision in any quantity and at any desirable distance and depth. Manure depositing machinery has, heretofore, opposed the greatest difficulties, but these have rapidly yielded to experience and skill, and now even damp compost is put into the soil conjointly with the seed, and with a regularity which leaves little room for further improvement.

By this time two types of seed drilling machines had become firmly established, the force-feed and the cup-feed, both of which were almost identical in outward appearance. Only their seed delivery mechanisms differed, the former making use of small brass fluted cylinders, the latter of cups, to take up the seed and cast them into funnels.

Many drills were provided with a large hopper to carry dry manure, which was sown along with the seed. It was dropped by a separate mechanism and did not come into contact with the seeds until they reached the soil. There were also various patterns of **drill barrow** for sowing peas or beans one row at a time, a multitude of one and two row **turnip drills,** as well as **broadcast distributors** for grass and clover seed. By the middle of the century most manufacturers were including general purpose drills amongst their range. These machines were sturdy enough to withstand the hardest buffeting and were well adapted to all seeding requirements on the farm, being capable of drilling all kinds of

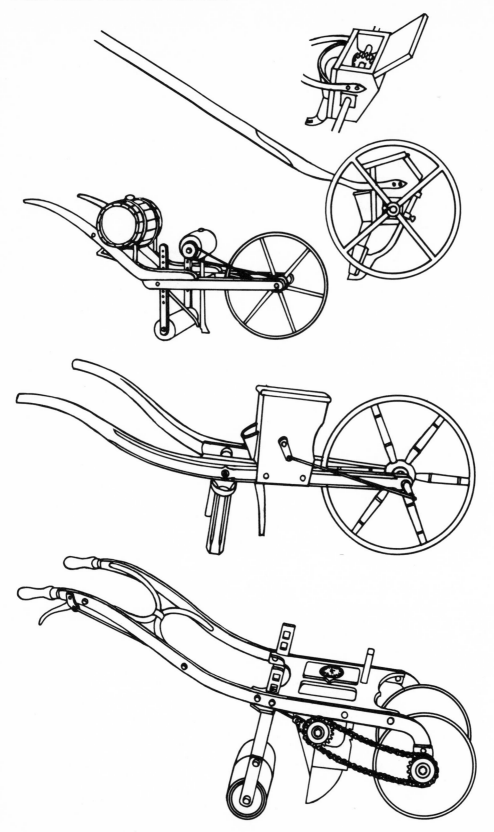

Single-row seed drills, with notched cylinder delivery

corn and seed, with or without manure. The arrangement of such a drill is shown on the accompanying illustration. The cup-feed mechanism and the manure barrel were driven by toothed gearing from the travelling wheels, the gears being situated on the side of the hopper. The seed delivery tubes were a form of continuous flexible tubes down which the seed was dropped to where the coulters had prepared small furrows for its reception.

A great number of different seed drills were exhibited at the British Agricultural Shows during the remainder of the century. Their use had become widespread in America and other parts of Europe where manufacturers could be found. But all these machines were variations on the principle already established by the early inventors, and they have changed very little since.

The broadcast-sowing machine appeared when English inventors and users of seed drills were but a handful. At that time opinions were very much divided upon the merits of broadcasting, dibbling and sowing by the new mechanical means. But more seed drills were gradually introduced and amongst them was the broadcast-sowing machine. It quickly became established in England and was introduced into Scotland about 1817, according to James Slight and R. Scott Burn, who quoted the reaction of the purchaser: 'The first broadcast sowing machine that came into this county (Berwickshire) was ordered by myself from England, from Mr Short of Chiverton by Hackhill: it was a short thing, wheeled by a man and was about 8 ft. wide.' R. L. Ardrey recorded that a number of these wheelbarrow drills were sent to America, where the manufacture of such items did not begin until about 1840. The early form, as suggested, was a wide, narrow, seed box, with a lid, mounted transversely across a light wheelbarrow framework. It was intended to sow small seeds broadcast, which it did in an efficient manner, with greater regularity and

precision than could be achieved by any sower's hand, and as its use increased, so did the length of the seed box until a convenient and manageable length of about 16 ft. became common. This dimension gave rise to the problem of transportation and so the seed box was divided into three sections, hinged together so that the two outer ones folded in over the centre section, making storage and movement through gateways much easier. Inside the seed box, the sowing mechanism consisted of a spindle about $\frac{1}{2}$ in. square which ran from one end of the box to the other. This was made to revolve by a chain and gear wheels or a vertical drive shaft from the barrow wheel, and at intervals along the axle were thin, radial iron teeth or stiff brushes. As the machine was pushed along by hand from the rear or pulled by a single horse, the long axle and the radial teeth revolved, the action of which agitated the seeds within the box and caused them to be thrown out through perforations along the rear side of the box. From these holes the seed fell in a constant shower, distributing themselves evenly to the full width of the machine. The size of the holes could be adjusted to regulate the amount of seed being sown by means of a long iron strip, positioned inside the box, which could be brought down partially or fully to cover the perforations. This barrow form of broadcast drill has been used until recently by which time its stability had been improved, in some models, by the addition of a second, and even a third wheel, or a following roller. Some had a light harrow attached to trail at the rear. Wherever broadcast sowing is completed nowadays the drill is hauled at the rear of a tractor.

The drill barrow was the simplest of all machines devised for sowing seed. It was often constructed at home, on the lines of a wheelbarrow, for the purpose of sowing a single line of turnips, peas or beans etc. Widely used during the whole of the last century, they were so multifarious that no two

Broadcast-sowing machine for grass and clover

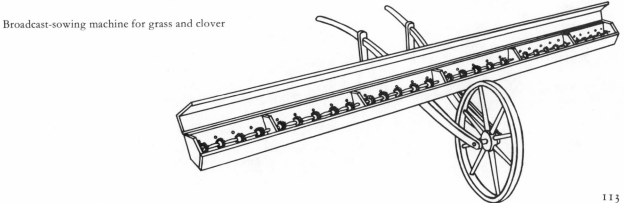

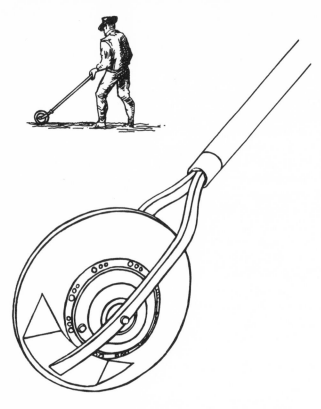

Late nineteenth century seed drill for single-row dropping, manufactured by A. W. Gower & Son, Market Drayton

barrows appear quite the same, although each had a hopper to contain several pounds of seeds, a device for dropping the seeds and a funnel or coulter to deliver the seeds into the soil. Various barrows are illustrated. They have not been much used since the turn of the century, except perhaps by the isolated farmer with a small extent of land.

Hand seed drills such as the one illustrated were useful for filling in blanks or misses where corn, turnip, onion, cabbage and carrot seeds etc. had not germinated. This model made by A. W. Gower, of Market Drayton, during the nineteenth century had a pole handle 4 ft. 6 in. long by which the 9 in. diameter galvanised iron disc was pushed along the row. Seeds deposited in the small hopper at the centre of the disc fell out through holes in its perimeter as the disc and hopper rotated, being dropped into the soil at regular intervals until the motion was halted. Many variations of this useful machine were available for market gardeners and the like until very recently, but efficient seeding drills have now rendered them unnecessary.

Turnip seed drills replaced hand broadcast sowing in Scotland during the early nineteenth century but almost forty years were to pass before they were fully accepted in England. The illustrated turnip drill by Moodie and Co., is typical of the nineteenth-century form and similar models are still to be

Moodie's turnip drill, from the *Popular Encyclopedia*

found. The seeds were contained in a cylindrical vessel which rotated longitudinally. Being small and round they were able to move freely about inside the vessel, making their exit through holes of the appropriate size in its circumference and falling evenly across the land. Most turnip drills were designed to sow two parallel rows of seeds, in addition to which some deposited manure along with the seeds.

A drill for ploughing corn, peas, beans etc., was established in England during the last century. It worked on a similar principle to the drill barrow, but could not function without the assistance of a plough, to which it was attached. The apparatus

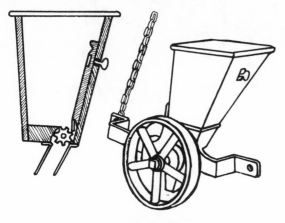

The corn sower was bolted onto the side of a plough

capable of holding several pounds of seeds was equally suited to single, double and multi-furrow ploughs, and simply hooked onto the side or between the stilts. The seed was dropped by the use of a fluted, rotating roller placed in the bottom of the hopper, and driven from a wheel travelling along the bottom of the furrow. The quantity of seed sown was regulated by a bristle brush pressing against it, but various rollers could be interchanged when sowing different kinds of seeds. The dropping mechanism could be halted at the headlands, or whenever necessary, by the ploughman pulling on a chain or lever to raise the wheel out of contact with the soil. This device was particularly useful for late wheat sowings, as a farmer could sow his land whilst ploughing, without additional labour. Similar to the drill barrow, this device was laid aside when tractor-drawn ploughs and seed drills were developed.

The soot distributing machine invented in 1839 by Scotsman Alexander Main was the first mech-

anical means of distributing dry manure. It could deal effectively with bone dust and bird guano, as well as soot and other powdered manures which had hitherto presented great difficulties, either blowing away on the wind or alternatively suffocating the worker who endeavoured to scatter by hand. Main built his machine as a broadcast distributor, employing for the purpose a wooden chest about 3 ft. deep and 6 ft. wide, carried upon an axle and two travelling wheels. The axle passed through the full width of the chest and carried a fluted roller 6 in. in diameter along its length. This roller was arranged to fit closely into a slit opening along the underside of the chest, whilst immediately above it, also enclosed in the chest, was a sheet iron cylinder 22 in. in diameter and 6 ft. long, mounted upon an axle and turned by gearing from the wheel axle. Soot was placed inside this cylinder through a trapdoor which was held shut by a catch while the cylinder revolved. Soot escaped through a multitude of tiny holes pierced through its surface, and fell in a continuous shower upon the roller below, where it was caught up in the flutings, then dropped onto the soil as the roller turned over. To maintain a constant flow of soot, a long bristle hair brush was placed in contact with the side of the fluted roller and its gentle but firm pressure wiped the turning roller clean, except for the soot contained in the deep flutings which passed by the brush and then fell down onto the surface of the land. This ensured equal distribution until the contents of the cylinder were exhausted, then the trap-door was opened and any foreign matter, such as rubble, mortar or stones, which particularly accompanied chimney soot, were removed before refilling. The action of this machine was later improved by the substitution of delivery wheels, set at intervals along the length of the axle, instead of the fluted roller. To accommodate this new feature, the bottom of the hopper was closed except for an apperture $1\frac{1}{2}$ in. in diameter below each delivery wheel.

Other broadcast distributors had evolved before the end of the nineteenth century, the most successful having a number of horizontal discs mounted side by side beneath a hopper on wheels. The discs were rotated through gears from the motion of the land wheels, and powdered manure, or the new chemical fertilizers, falling down over them in a constant but regulated trickle from the hopper above, was thrown out in all directions quite evenly. At the same time many farmers were

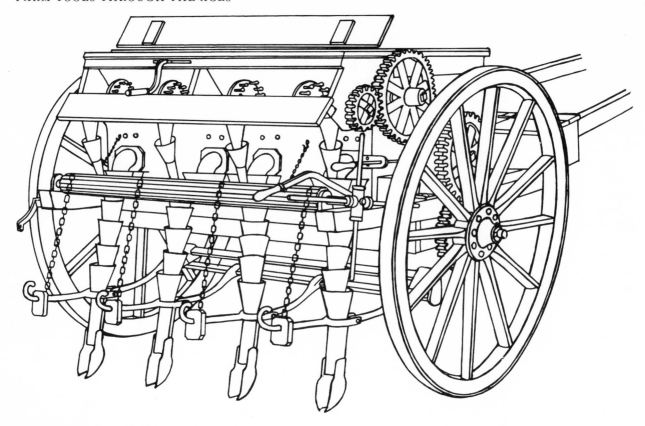

Chandler's liquid manure drill, c.1885, adapted for sowing corn and other seed together with liquid manure

managing by shovelling dry fertilizer over the back of a moving waggon. This was a very wasteful method.

The modern type of distributor for sowing fertilizer broadcast is based upon the last mentioned machine and nowadays consist of a brightly coloured plastic hopper, like an upturned cone, with a single disc below it. It is mounted on the rear of a tractor which supplies the drive, and manure falling from the hopper onto the rotating disc is scattered very evenly for a considerable distance.

The liquid manure drill was designed to deposit a measured quantity of liquid farmyard manure along with the seed. About the middle of the nineteenth century, Thomas Chandler of Aldburn devised a simple mechanism for delivering the manure that was incorporated in all later drills of this class. It consisted of small cups or dippers attached to an endless chain, rather like a dredging machine. As the drill went along, the chains rotated and the cups caught up the liquid manure from the tank, tipping it into the mouth of delivery pipes as

they turned over. A number of such units were mounted side by side behind a corresponding number of seed droppers, the seeds being contained in a separate hopper and delivered by the mechanism of a common drill. It would deliver liquid manure of any consistency without blocking and could also distribute the new dry compound fertilizers.

The collection and distribution of liquid organic manure does not have a place in modern farming, but combine seed drills with separate compartments for dry fertilizer and seed are used extensively. Farmyard manure or straw dung is no longer distributed over the fields from the rear of a horse-drawn cart as there are tractor-drawn machines, the commonest being a long narrow trailer from the rear of which dung is scattered in fragmented form by the action of rapidly revolving forks. The spraying of chemical fertilizer is now largely done by contractors who use various kinds of vehicle especially adapted for the purpose. Helicopters and aeroplanes are employed in those countries where the fields are larger.

Potato planting was done manually until after the middle of the nineteenth century. The earliest form of horse-drawn planter patented in 1857 dropped tubers at regular intervals by a form of rotary distributor. Mouldboards were provided at the fore and rear to open the soil and then close it immediately over the potatoes. Of the efficient planters that followed, the best consisted of two hoppers filled with potato sets and mounted upon travelling wheels. An endless chain formed with a series of cups was driven from the wheels and passed up through each hopper, each cup gathering a potato during the passage. The chains turned over pulleys and dropped the potatoes into tubes by which the chains themselves descended, and as each cup left the bottom of the tube, a potato was discharged into the furrow. A different mode of distribution had been devised by about 1870. Potatoes in the hopper were impaled one by one, on small forks attached to the perimeter of revolving discs, and dropped into the mouth of a delivery spout. They were directed into a furrow made by a wide coulter and covered over by following mouldboards. Spikes were replaced by a picker or pincer arrangement before the end of the century. In some instances, potatoes are still set in the furrows by hand, whilst there are efficient potato planting machines based upon the earliest patents. They are followed in the late autumn by equally efficient potato harvesting machines.

The liquid manure cart. In early times it was realised that both solid and liquid manure contained valuable plant foods, so consequently efforts were made to collect and distribute both in measured quantities over land where crops were grown. Solid manure was carried in carts and handled by suitable spades, rakes or forks, whilst liquid manure was collected in containers and then released in the desired areas. This practice existed in many countries, and the turn of the eighteenth century saw a variety of carts with barrels or tanks mounted upon them for distributing liquid manure in quantity. They were simply a development of the common, horse-drawn water cart which in some countries was used to transport drinking water, and in others to water the streets or moisten the land during drought. The liquid was released by means of one or more taps, or pipes, hanging over the rear of the cart, and gushed out in continuous streams through a multitude of tiny holes until the tank was empty. As the century progressed, farmers contrived to

improve this system of distribution and also to collect all liquid waste from outhouses, stables and cowsheds by directing it along gutters and drains to a dung heap where it was filtered through layers of straw. Any solid matter was retained by the straw whilst the liquid was collected in a large masonry vat constructed below ground level. In this way nothing of value was lost; even rain water that washed the farmyard, soap fat, grease and salt from the kitchens were collected and returned to the land. Manure carts were generally replaced, during the early part of the present century, by liquid manure drills which were considered less wasteful. The latter did not distribute the fertilizer broadcast but placed it where it was most needed in the furrow along with the seed.

A chain pump was installed on most farms during the late nineteenth century as a convenient means of raising liquid manure from an underground tank. It was larger and required more effort to work than the ordinary water pump but was more robust and not likely to block up or freeze during the winter. There was little difference between one make of chain pump and another, as they simply consisted of a long iron pipe some 6 in. in diameter, the bottom end of which was secured by a special fastening to the floor of the tank whilst the top extended up for several feet above ground level and terminated in a long, movable spout, through which the liquid was directed into a **manure cart**. At the point where the spout joined the pipe there was a large iron wheel which the operator reached by standing on an elevated platform built over the mouth of the tank.

Cart for liquid manure or water

Subsequent turning of the wheel caused an endless chain, with cast-iron discs spaced out along its length, to rise up from the bottom to the top of the pipe, forcing the liquid up with it. The discs were inserted into the chain at intervals of about 18 in., and they were the same diameter as the inside of the pipe so that each one acted as a platform and elevated a column of water, until it passed out through the spout. The chain made its exit at the top of the pipe, and after passing round the wheel was returned clear of the pipe to the bottom of the tank where it again entered the open bottom and made another ascent. Chain pumps continued to provide reliable service alongside the rotary diaphragm pumps introduced during the early twentieth century. However the chains and discs gradually rusted away and could not be replaced, but the remaining vertical pipe with its wheel and spout can still occasionally be seen *in situ*.

Hand Sowing

The sowing basket, as its name suggests, was made from woven wicker. Later versions however were made in wood or tin. The size and shape of the basket was determined to some degree by local tradition, but in all cases it was made to hang from the seedman's shoulder on a leather strap and fit closely to his side. The strap was fastened to the edge of the basket nearest the sower, and on the other edge there was a single handle or projecting stave with which to hold the basket steady. It was customary for one sower to complete an entire field so that the distribution of seed was even throughout. He would refill his basket from a large **seed rusky** positioned at the edge of the field. The use of the sowing basket persisted long after the seed drill was evolved but was not common in this present century.

The sowing basket

Wooden containers used in place of seed baskets

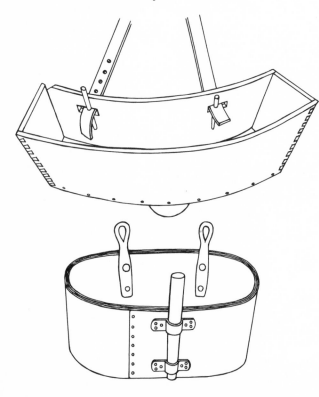

The double handled sowing basket. If the sower was accustomed to sowing seed with both hands then the pattern of his basket was such as to suit this manner. It was made of wicker, tin, or wood, and fitted against the front of his body, secured in position by a stout leather belt which passed round his waist. Further attachments, in the form of leather straps, were taken from the front and side edges of the basket and around the sower's neck so as to support the loaded basket adequately and leave his hands completely free to cast the seed. Neither did the use of this basket extend much into the present century.

The sowing sheet

The sowing sheet. In some districts, the broadcast sower used a large cotton sheet rather than a basket to contain his seeds. It was folded and draped around the top half of his body, and, as shown by the illustration, it then formed a deep bag, suitable for containing a large quantity of seed which he carried with ease whilst his right arm remained free for

scattering them. This was not a common method of carrying seed and was appropriately displaced by the seed basket sometime during the nineteenth century. **The seed fiddle** emerged during the eighteenth century and replaced hand-broadcasting to a limited degree. It was, as illustrated, a canvas bag and wooden frame of convenient size to be suspended by a strap around the sower's neck and rest against his chest. Some 7 lb. of seed, contained in the bag, trickled down in a regulated stream into the horizontal disc which rotated, to right and left alternately, through the action of a thonged bow. The movement of the disc scattered the seeds in a wide arc about the sower who walked at a steady pace in order to cover ground evenly. A similar sort of broadcasting instrument had the disc rotated by a small crank handle and a system of gear wheels instead of a thonged bow. Seed fiddles were sold until quite recently and used for sowing small amounts of grass and clover seeds.

Late nineteenth century fiddle sower

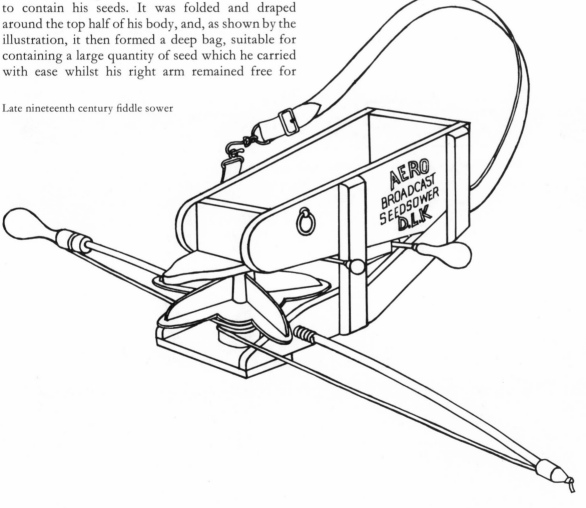

Seed Dibbling

The seed dibber. The application of dibbers to setting wheat was first proposed by Edward Maxey in his *A New Instruction of Ploughing and Setting of Corn*, 1601. He credits the discovery to 'some sillie wench having a few corns of wheate mixed with some other seed'. She was careless enough to drop them into regular spaced holes which resulted in a growth of corn more vigorous than if broadcast-sown. This gave some occasion for further trial, after which he recommended his setting board, 3 ft. long and 12 in. wide, pierced with regular holes about 3 in. apart. The worker knelt on the board and inserted a wooden dibber through each hole

The setting board working behind a plough. From Edward Maxey's *A New Instruction of Ploughing and Setting of Corn*, 1601

into the earth below for a depth of about 3 in., the penetration being regulated by wide shoulders on the dibber. Setting was kept straight by means of a gardener's line along which the board was moved at regular intervals, until the row was completed. Seeds were dropped into the holes by a second person whilst dibbling proceeded, then the soil was raked over lightly.

The practice of hand dibbling was praised again in 1803 by Arthur Young, who could not 'expatiate sufficiently on the excellency and importance of the dibbling husbandry'. It saved several pecks of seed per acre, afforded a second harvest to the poor, the grain commanded a higher price at market, but most important of all was its even germination and convenience of hand hoeing between the growing lines of plants. By the end of the

nineteenth century dibbling was keenly practised in eastern England, and few farmers there sowed their seed broadcast if they could get enough hands for dibbling.

The single dibber continued in use especially in Norfolk and Suffolk long after seed drills had become available. Their common pattern consisted of a wooden shaft about 3 ft. long, equipped at the top with a cross-handle. From the other end there extended an iron shank about $\frac{1}{2}$ in. square, which bulged out towards the end to form a dibbling point, about 2 in. in diameter and 3 in. in length. Often the whole instrument was forged by a blacksmith and called a setting iron or dibbler, though the latter term usually described a person doing the work.

The double dibber was made from two single dibbers and speeded up the work considerably. It was common practice to link two dibbers together by means of a thin iron plate about 9 in. long and 3 in. wide. This plate was pierced with a hole near to each end, and the shanks of the dibbers passed through these holes. It was retained by the handle and the dibbing points below. One dibber was held in each hand and pressure from one foot was exerted on the plate, thus forcing the dibbing points into the ground. Expert users could maintain a rapid walking movement with this instrument and by

Corn dibbers

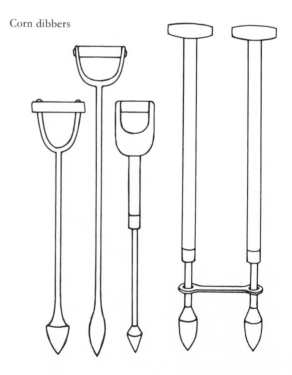

swinging one dibber in an arc around the other they left a straight row of holes, the same distance apart, ready to receive the seeds. Another method of double-dibbing was for a man to walk backwards with a dibber in each hand and with them strike two parallel rows of holes across the width of the field. A woman, usually his wife, followed behind and dropped a few grains of corn into each hole, after which it was covered by harrowing.

The hand dibber used in France and the Netherlands during the nineteenth century was called a *plantair* and consisted of two pointed wooden stakes, each 3 ft. in length. They were held apart some 12 in. by a wooden cross-piece half-way down the length. A second cross-piece was fixed across the top of both stakes and also joined them together, whilst extending out beyond each to form side handles. This instrument made two holes at once when the operator pressed the bottom cross-piece with his foot to drive the points into the soil. It was commonly employed for transplanting.

A common form of hand dibber used in the eastern counties of England prior to this century was comprised of several wooden or iron spikes each about 2 in. long, set at intervals of 2 or 3 in. into the underside of a wooden board. There was a staff or pole handle attached to the top side so that the spikes could be pushed down into the soil by pressure from the user's arms, whereupon it immediately made as many holes as there were dibbers. Sometimes the spikes were tipped with iron or were wholly of iron, and some English eighteenth-century corn dibbers were hollow so that a core of earth was ejected through the top as work proceeded. The person who used the dibber would work in line with a rope, stretched taut across the field, and he would be followed by a woman sower who dropped seeds into each hole. A **hand rake** or **bush harrow** was then used to cover the seeds lightly.

A form of dibber common in Suffolk had a wooden cross-head fitted with a number of iron spikes. The user pressed them into the soil with his feet or by the pole handle, and women or children following behind dropped a single pea or grain of corn into each hole until the field was fully set. The dibbling of corn lost favour as sowing machines were perfected.

Potato dibbers were formed in various ways for the purpose of making holes in the soil to receive the sets. The illustrated dibbers range from an iron model, designed to extract a neat cylinder of soil of

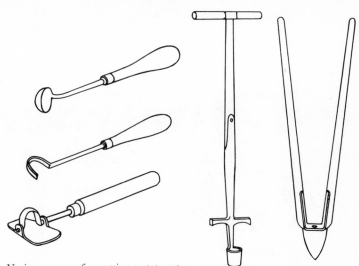

Various scoops for cutting potato sets
Potato dibbers were pushed down into the soil by foot-pressure

the correct size and depth for planting, to a crude but efficient tool grown naturally to shape in a hedgerow. Both were dependent upon the soil being damp enough to stick together, and were continued alongside potato planting machinery.

An improved seed dibber was still being manufactured by Boby's of Bury St. Edmunds at the end of the nineteenth century. It was made of tubular iron about $1\frac{1}{2}$ in. in diameter and 3 ft. in length with a wooden cross-handle at the top. The seeds were contained inside the tube and when the bottom end had been pushed down into the soil, the user extended one finger to pull a spring catch which released a number of seeds. The dropping mechanism was contained in the bottom of the tube and could be adjusted to suit the size and quantity of seeds to be sown. This was probably the last of the hand dibbing devices.

The dibbling wheel was an attempt to speed up the process, and some different forms were brought into use for planting potato sets, beans, peas, mangels, etc., but along with hand dibbers they were replaced by efficient drills for sowing the seed during the nineteenth century. The dibbling wheel generally took the form of a wooden wheel about 3 ft. in diameter, contained at the fore-end of a barrow-like framework with handles, by which the operator pushed it along and upon which he placed any extra weight thought necessary. A number of iron dibbling points, each about 2 in. in length and $1\frac{1}{2}$ in. in diameter, were attached around the circumference of the wheel, and the interval between the points was adjusted to suit the type of seed being sown. As the wheel revolved the points entered and then left the soil, leaving a succession of holes into which a following person deposited two or three seeds, afterwards closing the holes with his foot or by raking. The biggest disadvantage with a single wheel was that the operator was treading in a direct line with the holes left by the points, which would either mis-shape or destroy them completely.

About the middle of the nineteenth century improved versions with two wheels set side by side were discussed in the farming journals, but they were too late, owing to the increasing popularity of seed drilling.

Tree Planting

Trees were planted on a large scale by gentlemen farmers who owned sufficient acres to make timber cultivation a worth-while proposition. During the eighteenth and nineteenth centuries there were several modes of tree planting, each related to the terrain and the number and size of the trees to be planted. **Slit planting** was the simplest method and was appropriate for transplanting small trees where the earth was loose and free from top growth. It required no prior preparation in the manner of deep digging, ploughing or trenching, and was performed by a man using one of the following three hand instruments: the moor planter, the diamond dibber or the common garden spade. On some occasions he would be assisted by a second person, usually a boy, who carried the seedling trees in a basket or canvas bag.

The moor planter was a type of dibber and consisted of a wooden helve 2 ft. 9 in. in length and $2\frac{1}{2}$ in. in diameter, which terminated with a slightly curved iron prong, 15 in. long. The workman grasped the planter in both hands and thrust its point down into the heath, and by depressing the helve, the prong forced up the soil, leaving a hole into which the boy placed the roots of a seedling tree. He then compressed the loose soil around the roots with his foot whilst another hole was being prepared.

The diamond dibber was used for tree planting where the soil was gritty or stony and the surface growth sparse. It was forged with a triangular blade of iron, each side being 4 in. in length. The blade was furnished with a short iron shank and wooden cross-handle, and the total length of the instrument was about 12 in. Its small dimension allowed the user to work unaided; he held the dibber in one hand and supplied his own trees from a basket strapped to his waist with the other hand. To make a suitable bed for the root he angled the blade towards himself and worked it down into the gravelly soil, then pulling up the handle he formed an opening which was kept open by the blade whilst he placed the tree in position. The dibber was afterwards removed and the soil made firm by trampling.

Slit planting was practised with a common garden spade whenever it was necessary to cut through turf, which was incised and drawn back before the hole was completed. These implements were much used during the eighteenth and nineteenth centuries, when trees were planted on a larger scale than ever before. Mechanical equipment nowadays executes the same work at a much faster rate.

The planter's mattock was an absolute necessity in rough, rocky districts where soil was thin and heavily infested with roots. It was stoutly built, with a wooden handle 3 ft. 6 in. in length, and an iron blade formed on one side into a pick, 17 in. in length with the other side being 16 in. in length and finished with a cutting edge 5 in. broad. Both the pick and the cutting edge were faced with steel and kept very sharp in order to equal the task of cutting through tough roots and reducing soil to a suitable condition for the planter's spade.

The planter's spade was little more than a scoop or hand-hoe. Its slightly concave iron blade was attached at right angles to the end of a wooden handle about 12 in. in length, and the user scraped or scooped the loosened soil, fragments of stone and other debris from the hole as he excavated it with the mattock.

Hackle prongs were used by the tree planter instead of a digging spade when the ground was sticky or largely composed of clay. They were similar in size to **trenching forks** and, in like manner, were made with both two or three prongs, the lesser number being suitable for clay and the larger number, three prongs, when the soil was full of gravel or flints. Planting mattocks, prongs, forks and special tools were locally made by blacksmiths until recently. Tree planting is mostly completed by spade only nowadays, but mechanical means are available should the terrain require it.

Birds and Insects

Bird scaring. Country boys could usually earn a few pence at sowing and harvest time by frightening away birds from the farmers' fields. But the seemingly simple work of 'bird hollering' involved long hours of yelling, yodelling, ringing handbells, swinging wooden rattles or creating a noise by some means or other.

Egyptian and Roman boys no doubt did the same thing many centuries ago, perched on high platforms or in the branches of a tree where the sound of their voices or the shaking of pebbles inside a jar would carry a good distance.

In later times boys usually provided themselves with a pair of clappers or a rattle, the former being simple instruments that could be made by the boys themselves, whilst the rattles were purchased from a carpenter.

Clappers were held in either hand and consisted of an oblong bat of wood about 12 in. long and 4 in. wide with a 4 in. square of equally hard wood on either side. The three pieces were loosely connected together by a loop of leather thong, and shaking the clappers caused it to emit a loud and continuous noise.

Wooden rattles give out an ear shattering clatter when they are rotated vigorously. They have been

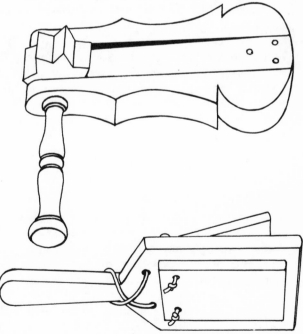

Clappers and rattle for bird scaring

123

employed for sporting rather than agricultural purposes during this present century and during World War I were used to forewarn soldiers of a gas attack.

The humble scarecrow came under the scrutiny of mechanically-minded men during the nineteenth century, some of whom designed mechanical alarms to replace it. But the ragged scarecrow, however inefficient it might have been, cost nothing to make or maintain whilst the new alarms invariably depended upon delicate mechanisms, used large quantities of gunpowder and required daily, if not more frequent attention.

The rook battery, as sold before 1900 by Messrs Slight of Edinburgh, was one such contrivance intended to frighten away the largest of birds from growing crops. The battery was arranged to stand upon a tripod at one side of a field and to emit a series of loud bangs throughout the hours of daylight. It consisted of a circular tin chamber 18 in. in diameter and 3 in. in depth with a detachable conical cover. There were twenty-four apertures each $\frac{3}{4}$ in. square pierced through the wall of a chamber and from each projected the muzzle of a 4 in. long brass cannon. The cannon radiated outward from the centre of the chamber and were attached to the base by catches, so that they could be removed for the purpose of recharging with wadding and gunpowder. A match thread was attached to the touch hole of each cannon and as it burned down caused each cannon to explode in turn. The makers recommended that the battery be moved to a different part of the field each day to ensure that the rooks did not become too familiar with it. Not a great many of these noisy affairs were ever purchased, but the use of frequent banging to simulate gun shots was later taken up by other makers whose alarms made use of mechanical or clockwork movement to fire a series of large percussion caps. They banged away right through the night and probably disturbed the countryman's sleep more than they did rabbits and hares.

The Paul net. The idea of capturing harmful insects, especially the turnip fly, received some attention in the early years of the nineteenth century. Farmers were recommended to sow a small patch of white turnips as a decoy, some time before the main crop was put in. The insects attracted to this early crop were removed by sweeping with a net and therefore subsequent generations were destroyed. Alternatively a freshly tarred or painted board was

rapidly drawn along the rows of turnips thus causing the insects to jump up from the leaves onto its sticky surface from which they could not release themselves. An extension of this idea was patented by a Mr Morris about 1840 and his ingenious mechanical device called a turnip fly catcher, was manufactured by Messrs Banks and Nixon of Nottingham.

The turnip fly catcher involved the use of a small two-wheeled carriage which was propelled by means of two handles along a drill of turnip plants. Two wheels supported a light iron framework to which the handles were an extension. The base of the framework was made horizontal to the surface of the soil, and could be adjusted to a convenient height for brushing the leaves of growing plants. A canvas sheet, heavily smeared with adhesive substance, was attached to the underside of the frame so that when the device was pushed forward over the turnip plants, the canvas flapped against the leaves and caused the disturbed flies to be caught up and held by the adhesive. Such devices were the preoccupation of Victorian inventors but were hardly practical.

5

Harvesting the crops

Reaping Machinery

Amongst the outstanding developments of the nineteenth century was the gradual perfection of a corn reaping machine and its subsequent development into the reaping and binding machine. There were some who doubted whether standing crops could ever be cut down by anything but sickles and scythes, but the more enlightened agriculturalists were aware that a machine for the purpose would be of enormous benefit to corn production. There was a tendency to sneer at the very notion of it ever becoming reality and, as the *Farmers Magazine* of 1806 was anxious to point out, 'those who remember the general opinion when threshing machines were first introduced, will not be sceptical concerning the success of a machine for reaping corn. The latter will no doubt require many years and many alterations before it can be brought into complete and general use.' In 1783, The Royal Society of Arts offered a prize for machinery 'by which the mowing or reaping of wheat, rye, barley, oats or beans may be done more expeditiously, and cheaper,' but no practical machine came forward and indeed the corn harvest continued to be reaped almost wholly with sickles and scythes for almost another hundred years.

Joseph Boyle of London secured the first patent for a reaping machine in 1800. He incorporated a circular iron plate fixed horizontally beneath a carriage frame and rotated as fast as possible through gearing from the travelling wheels. The revolver had a number of scythe blades attached to its perimeter for the purpose of cutting the corn, but there was no device to hold the stalks onto the cutter. Neither was there any provision for collecting the corn, and it is not known whether the machine was pushed or pulled into the crop. This inventor was striving to reproduce the action of a scythe as used by the hands, an obsession which lingered with other inventors for some time after.

Thomas James Plucknett, an implement manufacturer of Blackfriars Road, London, was next to use a horizontal rotating disc. This time it was sharpened and notched around its edge to cut the corn like a circular saw. He mounted it at the fore-end of a carriage frame where it could be raised or lowered to suit the height of the crop, which was directed onto the cutter by a comb. But unfortunately the cut corn was scattered all about as there was no provision for laying it over.

Gladstone, a millwright of Castle Douglas, Kirkcudbright, used a saw-edged circular cutter driven by a belt and gearing from the travelling wheels of the machine. He included a comb with long forward pointing teeth to gather the crop, and rotary forks that turned in the same axis as the cutter to secure the stalks against the comb whilst they were cut through. He gained the interest of the Highland Agricultural Society with a model of his machine, whereupon they encouraged him to build it to full size. But for some reason the Society lost interest, and the project was discarded. Continuous rotation did not answer, so the next stop was to imitate the action of hand shears or clippers and cut the stalks with short successive strokes. John Salmon of Woburn, a man whose ingenuity resulted in various other agricultural implements, built a machine with cutters based upon a scissor action. He also devised an apparatus for collecting the cut corn and delivering it from the rear in bundles ready for binding. The idea raised great hopes, but as with former machines, it was soon laid aside through the lack of real encouragement. It would appear that Robert Meares, of Frome, Somerset, was ahead of John Salmon in using shears. He was earlier granted a patent for what was simply a huge pair of scissors carried on wheels and operated by one man.

A Northumberland farmer, Donald Cummings, who in 1811 patented a reaper with revolving knives, said the cost of harvesting with a sickle or hook was about 12s. per acre, whilst he estimated the cost of reaping with his machine at about 4s. per acre. He pointed out the calculated saving was 8s. per acre in labour costs alone, to which could be added the benefit of swifter, safer stacking and, as the machine cut closer to the ground, a far larger amount of straw.

In 1812 the Dalkieth Farming Club offered a premium of £500 for a successful reaping machine and surprisingly there was only the one entry, that of James Smith, the celebrated and eminent land drainer of Deanston. Here again the cutter was of circular form, but this was attached to, and projected out a little beyond, the base of a vertical drum 4 ft. in diameter. The necessary number of revolutions for cutting were obtained when the horses moved at $2\frac{1}{2}$ m.p.h., and by a simple arrangement of gearing the drum could be made to revolve from right to left or left to right. The machine was pushed into the standing crop by one horse, yoked in shafts at the rear, an unusual and awkward position for both animal and driver, but one that nevertheless was necessary to save the crop from being trodden flat. The only defect in Smith's reaper appears to have been the manner in which corn was laid down after being cut. It fell against the side of the revolving drum and should have been laid in regular rows, but the varying pace of the horse affected the rotary speed and consequently the corn was strewn about at all angles. However the Dalkieth judges considered it to be the most efficient reaping machine to have come under their observation and awarded it prize plate worth 50 guineas. The inventor proposed to enlarge the diameter of his cutter to 6 ft. and to provide the power of two horses, but he never brought this to a practical state. Some reports estimated his reaper to cut about one English acre per hour, with an added note that the blade required constant sharpening.

A machine for reaping and at the same time laying the corn in parcels or sheaves was invented in 1822, by Henry Ogle, schoolmaster at Remington, Northumberland. One year later, Mr Brown of Alnwick Ironfoundry advertised them for sale. Their unpatented machine was observed to work well with wheat and barley and estimated to cut fourteen acres per day on even land. It was the first reaper to be drawn by horses rather than

pushed by them from behind. The cutting apparatus, a horizontal bar with teeth 3 in. long, projected from the right side of the carriage, and by motion of the travelling wheels, a steel knife was made to move rapidly from left to right cutting the corn trapped between the teeth. It seems to have embodied the principle of the later reapers, especially the American McCormick, even down to the revolving sails (Reel) which collected and pressed the corn into the cutting knife. On being cut, the corn was carried back by the sails onto a platform behind the cutter from where it was raked by hand into sheaves. The Ogle and Brown reaping machine embodied all the principles of a practical reaper, yet nothing more was heard of it or the inventors.

Joseph Mann of Raby, Cumberland, tried his reaper in 1822. He was sponsored by the Abbey Home Agricultural Society, who appear to have hindered rather than assisted his project to the point where the inventor abandoned it entirely. He later recommenced work and tried his improved version in 1826. His machine was unusual in appearance but it cut the corn well, making use of a revolving polygonal cutter with twelve equal sides and revolving rakes to gather the corn, which was laid in a neat and regular swath. Mann's reaper was drawn along by one horse attached to the fore-end and cut to a breadth of 3 ft. 6 in. It was demonstrated in a field of oats before the Highland Agricultural

Bell's reaping machine, as improved by Crosskill

Society's Show in 1832, and managed to complete an acre every hour, but to the disconsolation of the inventor, did not receive any reward.

A practical reaping machine continued as a matter of the greatest national importance, but the problems seemed insurmountable. Various soils and situations were enough to contend with, not to mention tangled and flattened crops which themselves presented much difficulty to hand reapers using sickles or scythes. All reaping machines constructed during the first half of the nineteenth century had a very limited life, if they ever worked at all, except however one designed in 1826 by Patrick Bell, who was minister to a small parish in his home county of Angus. By the time he began to construct his reaper he was familiar with the engraving of James Smith's machine, but he did not try to imitate or improve upon its working principle; instead he took his idea for the cutters from a pair of common garden shears, in much the same way that John Salmon had done years before. But Bell arranged the shears in an ingenious manner. He used a row of thirteen triangular blades sharpened on both side edges and firmly secured to a horizontal bar on the front of the machine at a height of a few inches above ground. Under and between the stationary blades were twelve reciprocating blades, the edges of which cut right and left with those in the row above. Thus a cutting action was provided when the sharp edges of upper and lower blades came together, everything in their path being mown down. The machine was pushed forward by two horses yoked behind to a pole and as it advanced into the crop corn was brought and held fast into the cutters by a revolving reel, very much like that on a modern combine harvester. The corn was severed near to the roots and then thrown back by the reel onto a travelling canvas which revolved to one side and deposited the corn regularly on the ground. The illustration of Bell's reaper omits the travelling canvas, since it would obscure the lower mechanical parts, but its direction is indicated by a curved arrow. This reaper was trialed secretly behind the closed doors of a large barn where Bell used some stalks of corn planted one by one on the earth floor. The experiment proved successful and a second trial was held at night time in a field during the autumn of 1828. The machine again worked well and was then demonstrated at Greystones Farm, in the parish of Monikie, before fifty or more gentlemen who

considered it most suited to the task.

Bell's original design underwent many alterations over the following years. He had stubbornly refused to take out a patent in order that a reaping machine might be brought into general use as quickly as possible, but with no patent protection he was unable to dispute the modification that any blacksmith or carpenter cared to introduce. Skilled and unskilled alike were free to produce their own versions, and of the machines that followed, only a few lasted more than one season.

About this time, farm labourers began to show their resentment against the new threshing machines and their mounting anger resulted in the revolt of 1830, largely through the fear that their traditional winter and wet weather work of flail-threshing in the barn would not be continued. Bell's original reaper was maintained, in the district of Inchmichael, Perthshire, where the Bell brothers reaped a harvest with it every year. Some of his machines were taken to the Continent, and some may have found their way to America where inventors had been working for some time towards an effective reaper.

It was two American models that caused a sensation at the Great Exhibition of 1851. They were the McCormick and Hussey machines. The cutter in McCormick's reaper was a low bar set with a row of stationary fingers. No clipping took place; instead the corn stalks, drawn towards the bar by a revolving reel, were held by the fingers and cut through just above ground level by a sharp oscillating edge which acted as a saw. The machine was drawn by horses and the cut corn raked clear of its platform and onto the ground by hand. Hussey's cutter consisted of a series of triangular blades that moved between two lines of projecting guards. The latter held the corn stalks quite firmly whilst the reciprocating blades cut them through. This machine was also dragged by horses and a man sat on the box covering the wheels and raked the corn from the platform by hand as it fell back from the cutter.

The question of whether Patrick Bell or Cyrus McCormick was responsible for the first practical reaper became a matter of debate. The story behind McCormick's success is equally as romantic as the one surrounding Bell. His father was a farmer in the county of Rockbridge, Virginia, U.S.A., who after experimenting with a machine in 1816 seemingly lost interest. He did however make

McCormick's original reaper

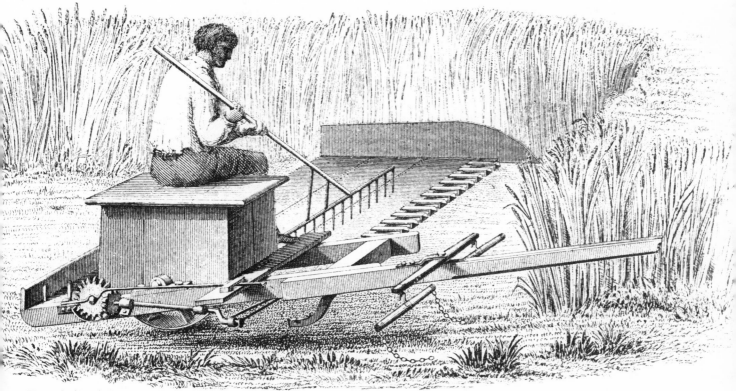

Hussey's American reaper

another unsuccessful attempt in 1831. At that point his son, Cyrus, became involved and was soon able to try out a reaper at harvest time amongst the sickles and scythes which at that time seemed irreplaceable. The machine did show some promise but the family had little money to spare for development work, the majority of which was carried out under exacting conditions in a small blacksmith's shop near Steeles Tavern, Virginia. The first McCormick reaper was sold in 1840 and within three years some fifty more had been dispatched. The demand quickly became so great that a factory had to be erected so as to manufacture the machine to order. McCormick left the problems of management in the hands of his brothers and proceeded to apply his ingenuity to promoting and improving the reaper, during which time he travelled considerable distances in order to study them in operation.

Reaper development reached its turning point and assumed a far greater significance with the 1851 Exhibition in London. McCormick attended personally to see his invention gain the highest awards. British agriculturalists were obliged to acknowledge its merits and Hussey's machine was equally appreciated, some preferring it to McCormick's. This major success enabled the Americans to open a market in Britain and then on the Continent, where different firms contracted to build the machine under licence. It was said that no fewer than 1,500 of Hussey's reapers were distributed in England alone, within two years of the exhibition. This competition had the effect of spurring British manufacturers on to such an extent that ten new reapers were shown at the Lewes 'Royal' in the following year. However, Patrick Bell's machine reappeared to challenge the Americans, and incorporating some minor but important modifications it defeated the Hussey-Crosskill reaper at the Highland Society's Show in 1852, and also defeated the McCormick machine a year or so later. But the Americans triumphed in the long run because of simplicity, and British manufacturers continued to build their versions of Hussey's and McCormick's under licence. McCormick's reaper was selling at £30 in 1852, with a royalty of £7 going to the inventor on the sale of each machine, and long before the end of the century his Chicago factory was producing thousands of reapers every week. Patents continued to be taken out at a lively pace in Britain and America, but interest remained centred on Bell's, McCormick's

and Hussey's reapers and the derivations of them. The harvest scene of 1859 was still very much unchanged, with scythemen doing most of the work.

The McCormick reaper had been further improved in 1854 with the addition of a Burgess and Key delivery system to replace the hand raking of cut corn from the platform. The improvement was a series of archemedean screw rollers to receive the corn from the cutters and deliver it in a neat continuous swathe at one side of the moving machine. Soon after, Ransome's brought out their Automaton reaping machine which, like the newly introduced Atkinson reaper from America, employed a circular rake to sweep the corn to one side. Other makers followed suit and this was to remain the accepted method for almost half a century.

Further trials continued to be carried out at the annual shows and the list of British, Continental and American manufacturers grew increasingly longer. It became necessary to divide the machines into three classes: side delivery, those without side delivery, and combined reaper-grass mowers. The last combination was never perfected and **mowing machines** became a separate item. At the 1869 Manchester 'Royal' a record number of eighty-four reapers were trialed. The established manufacturers, Bamlett, Hornsby, and Howards etc., each showed several machines and there were several newcomers.

But about this time, the prize system came up against a deal of criticism 'for dissipating the energy of manufacturers in pointless rivalry and frivolous competitions'.

British manufacturers owed a great deal to their American counterparts, and it was in America that inventors first turned their attention to furnishing the reaper with sheaf binding mechanism. The progress had already been made by the two Illinois farmers, C. W. and W. W. Marsh, who patented the first successful harvester in 1858 (the term 'harvester' is an Americanism applied to a machine that does more than simply reap corn). The Marsh brothers used a moving canvas to deliver cut corn over the centre drive wheel to a raised deck, where it was taken and tied into sheaves by two men riding upon the machine. The distinctive feature of their harvester was the steeply pitched canvas and raised table, both of which can be observed in later machines even when the men had been replaced by automatic binding and knotting mechanism. The illustrated McCormick harvester of 1880 bears much resemblance.

McCormick hand-binding harvester at work, *c.*1880

Walter A. Wood's 1876 harvester, with self-binding mechanism

The **mechanical sheaf binder** was the final stage in harvester development. An American, Walter A. Wood, patented a wire sheaf-binder in 1871, which was exhibited in England some five years later. A twine sheaf-binder had been patented as early as 1858 by fellow countryman, John F. Appleby, but it did not become popular because twine was so expensive. It reappeared in 1880, by which time twine had become the cheaper material, but it was doubly welcomed because short lengths of stray binding wire were found to be causing grave damage to animals and machinery. The Appleby twine binder and knotting device was so efficient that it remained the standard model.

The working of a harvester was briefly as follows:
1. The reel held the crop against the knife where it was cut, then swept onto the platform.
2. A travelling canvas carried the cut corn to one side where it was received onto
3. An elevating canvas, and raised over the centre wheel to the binding deck.
4. There it was slipped down onto a length of binding twine, and
5. It was compressed by two or more packing arms until enough had collected to form a sheaf, the pressure of which
6. Tripped a lever and motivated the binding mechanism.
7. A curved needle advanced and carried the twine around the sheaf, across the knotter bill and into the twine receiver, whereupon
8. The twine was knotted and cut through before
9. Ejector arms cast the sheaf over the rear end of the mowing machine.

Tractor-drawn harvesters that combined the McCormick cutter, the Marsh collector and the Appleby binder were in almost universal use till the 1940s. They are nowadays used for reaping where the climate does not allow combines to operate, and one still hears of the old reapers that are occasionally brought out to tackle a badly laid or tangled crop. Development of the combine harvester, in which harvesting and threshing are combined, commenced with Samuel Lane's U.S. patent in 1828. The first practical combine came eight years later from H. Moore and J. Hascall of Michigan. It had reaping and threshing mechanism, a winnowing fan and sack filling apparatus, but it did not work at first because of the unsuitable climate conditions. It was taken to California and performed successfully in 1854. The early combines were of course drawn by horses; steam and oil tractors completed the work at a later date. The self-powered combine as we know it today was introduced into England in 1928, where farmers etc. were not enthusiastic since the tractors, brought in earlier from America, had failed in the wet conditions. Such fears however were soon dispelled, as the provision of internal power resulted in an efficient manœuvrable machine capable of harvesting not only corn but a wide variety of standing crops.

The header harvester merely nipped off the heads of growing corn and elevated them to the body of the machine. The Romans had such a substitute for hand reaping. It was a form of low cart pushed into the standing corn from behind by oxen. An arrangement of long, closely-set teeth along the fore-edge combed through the corn stalks and ripped off the heads, which rolled back into the cart. The stalks were left standing whilst the corn heads were probably stacked for threshing at a later date.

A very similar arrangement was thus described by John Loudon in his 1836 *Encyclopedia*: 'A machine for reaping the heads or seed-pods of clover where the second growth of that crop is left to stand has been used in some parts of Suffolk and Norfolk. It consists of a comb, the teeth of which are lancet shaped, very sharp and set close. The comb is fixed horizontally along the foredge of the bottom of an open box, which is drawn by one horse and guided by one man, who empties the seeds in regular lines across the field.'

Much of the development towards a really efficient header took place in Australia, where harvest time always caused great anxiety owing to the shortage of man-power. The first Australian grain header, or 'stripper' was the work of John W. Bull and John Ridley in 1843. It was similar in principle to the Roman cart, but used rotating beater bars to knock off the corn heads whilst the stalks were gripped in the teeth of the comb. Later Bull improved the gathering comb and succeeded in threshing the grain directly from the standing crop. Another Australian, Hugh Victor McKay, successfully combined the stripper and the **winnowing machine** in 1884. Use of the header continued in the large grain-producing areas of America and Australia until it was superseded by the combined harvester in the twentieth century.

The harvesting of corn and hay overlapped in

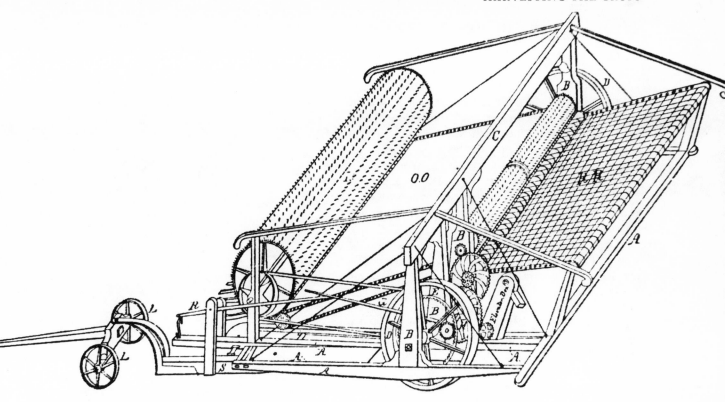

Machine for harvesting, threshing, winnowing and bagging
grain. U.S. Patent, 1836

that some tools were common to both operations.
The making of hay was primarily completed by
hand labour and required a complicated procedure
for drying, which briefly consisted of turning the
crop with hay-forks, raking it into single windrows,
building the windrows into grasscocks, the grasscocks into staddles, then into double windrows,
and finally building into bastard cocks. If the
weather was fine, the dry hay was then carried to the
stack. It normally required three or four days to
complete this preparation and employed about
twenty haymakers to every four scythemen.

Hand Tools connected with the Harvest

The ancient Egyptians harvested both wild and cultivated corn with a primitive form of sickle. In the first place it was made from the jawbone of a large animal, with the teeth sharpened down to form a cutting edge. Later the blade was fashioned in hard wood made slightly curved, edged with flints and very often set into a handle of baked clay. The Stone Age farmers had sickles made of flint, and the Romans used a one-handed instrument with the blade set at right angles to the handle. This was probably the earliest form of scythe. By the Middle Ages a curved iron sickle not unlike the modern one had evolved and was widely used for reaping corn, whilst hay was mown with an instrument very similar to the modern scythe in that it consisted of a long, slightly curved iron blade attached to the end of a pole that stood nearly as high as the user.

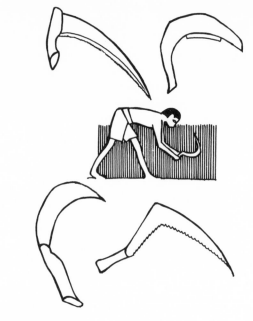

Primitive sickles

The reapers usually worked in teams of six with a selected man at the head. It was quite common for women to be amongst the reapers, but they were more frequently employed as gatherers who followed behind and laid the cut corn very precisely onto straw bands. When sufficient corn had been gathered to form a sheaf it was tied round with the band, and the sheaves then stood upright on their butt ends in stooks, or shocks of twelve each for wheat and ten for oats and barley. They remained there until dry enough for carting to the stack yard. Wheat could often be moved in about three days if the weather was good, the other crops remained in the field a little while longer. The **cradle scythe** was increasingly used for the purpose of cutting corn during the nineteenth century, and the sickle lost some of its importance. However both sickle and scythe were relegated to 'odd jobs' about the farm as **reaping machinery** gained in favour.

The sickle or reaping hook was used in one hand to cut down the amount of corn that could be grasped in the other. It had a curved steel blade with a cutting edge about 18 in. long and set with tiny teeth. The short wooden handle varied in shape throughout the country.

The cutting or bagging hook was less curved than the sickle but was broader across the blade. It was ground down to a thin sharp cutting edge which was used to hack over the corn in a manner called bagging. The advantage of this process was that it

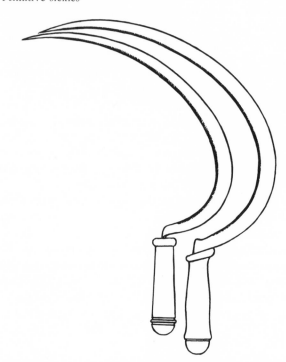

Toothed sickles

speeded the harvest, because the hook did not have to be drawn through the stalks but was struck against them whilst they were bent over by the free arm. Reaping was practised when the corn stood upright, and bagging was convenient when the crop had become laid or tangled by rough weather.

134

The reaping or cradle scythe. In many areas the scythe totally replaced the sickle for reaping corn during the nineteenth century. Its patent steel blade, strengthened with an iron strip along the back edge, was some 40 to 48 in. in length and straight except for the last few inches nearest the tip, which were curved. The blade was joined to a bent snead by a hook, ring and iron wedges, and like the mowing

Cradle scythe

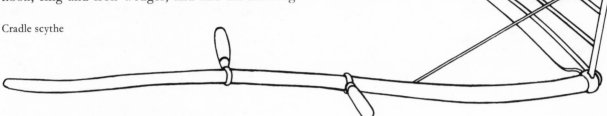

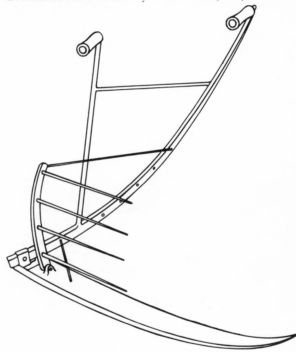

Drummond's iron harvest scythe with cradle, c.1850

scythe had a grass nail to hold it secure. A bent snead was normally preferred because it allowed the user greater command when working.

An important and distinctive feature of the corn reaping scythe was the cradle apparatus attached near the heel of the blade, and so formed to gather the corn and lay it down in a regular swathe across the field. Without the cradle, cut corn would have fallen in all directions and caused great inconvenience and loss of time to the gatherers who followed behind the reaper, forming the corn into sheaves.

The cradle attachment was fastened to the snead with the same iron ring and wedges which fastened together snead and blade, or sometimes was attached by wire through a hole near the heel of the blade. The dimensions and construction of cradles were exceedingly various. They were made mostly of wood, but often wicker and tin versions were employed. A popular nineteenth-century pattern stood about 14 in. high and had three horizontal wooden spikes, diminishing in length from the uppermost to the lowermost, set about 4 in. apart in parallel to the blade. A small cradle served well in England and Scotland but in America, Australia and other places where the crop was of a different nature, a larger cradle with more spikes was employed. In most, a hinge was located near to the point where the cradle joined the blade so that the whole cradle could be adjusted to the correct angle.

The Hainault scythe originated in Flanders during the Middle Ages and was used in that region to cut corn until the present century. Its use was observed by English travellers, who recommended it as a useful instrument in the hands of an experienced man, capable of cutting one-third more corn per

day than the English **reaping hook**. Consequently, Hainault scythes were soon introduced into England and were presented as gifts from the farmers to their labourers. This must have surprised the labourers, who as a rule were obliged to provide and repair their own harvest tools. However, this foreign scythe was not a success due mainly to the awkward action required for its operation, and the reapers quickly reverted to their hooks. The Hainault scythe had a short snead, approximately $1\frac{1}{2}$ in. in diameter, which was bent to an angle of

135

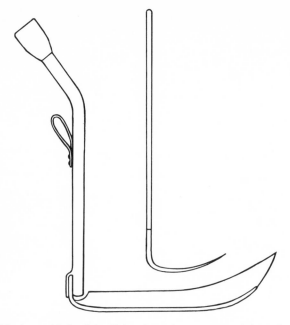

The heavy blade of the Hainault scythe was counterbalanced by a weight on the top of the snead. The handle of the hook was 3½ ft. in length

65 degrees near the top. The cutting blade was heavy, being 20 in. in length and some 3 in. in width across the middle. The Flemish method of using this scythe was for the labourer to grip the top portion of the snead with his right hand, having first inserted it through a leather loop to prevent the scythe slipping from his grasp. In his left hand he held a light staff tipped with an iron crook, with which he collected the standing corn and laid it over, and swinging the scythe with his right hand, he cut the corn close to its roots. The labourer proceeded in a straight line across the field, leaving the corn that he had cut leaning against the uncut corn. He would then return along the line, gathering the corn into sheaves by means of the crook which in turn he laid down to await the binders.

The mowing scythe was used mainly for cutting grass and took the simplest form of all the scythes. To form the snead, willow wood with the correct curvature was selected and, after being trimmed and tapered towards the top and perhaps even steamed into better shape, it measured between 5 and 7 ft. in length. The length and the shape of sneads was always a matter for dispute; consequently a good deal of variation took place from one area to another, but the individual could nearly always find one that appealed to him amongst the iron-

monger's or blacksmith's stock. Whatever the shape of the snead, short wooden pieces, shaped to fit the hands, were fixed at right angles to it by means of iron rings. They were usually placed about 18 in. apart, but their position could be adjusted to suit the user. A third iron ring at the base of the snead held the scythe blade in position, the blade being attached to it by means of a hook forged at the heel. The angle at which the blade came away from the snead had to be adjusted in accordance to the height of the user and the nature of the crop being mown. This was normally achieved by packing wedges between the snead and the iron ring. To ensure that the blade remained firm and retained its correct position during work, one end of a thin iron stay called a grass nail was hooked through a hole provided near the heel of the blade, and its other end secured to the snead. The grass nail also served the purpose of preventing any grass, thistles or weeds from lodging in the angle between the blade and the snead. Early scythe blades were hand forged to the specification of the customer, and were anything up to 5 ft. 6 in. in length by about 3 in. wide. They were forged from one solid piece of soft iron, that would bend rather than break when it met a stone. The nineteenth-century, factory-made scythe blades were considered inferior, as they mostly consisted of steel plate reinforced with an iron strip along the rear edge. Patterns of blades were however various and known as 'patent', 'labelled', 'crown' and 'extra warranted'. Whilst straight sneads were not common, they were supplied, larch wood being favoured for this form.

The scythe was widely employed before the present century for mowing grass and reaping corn; its use required a great deal of skill, and good scythemen were in great demand at harvest-time. However, since the introduction of mowing and reaping machinery it has only been used for menial tasks such as cutting down nettles and weeds about the farmyard.

The fagging stick was similar to the crooked staff used with the **Hainault scythe** in Normandy, and it probably existed for some centuries in England and Scotland before that scythe and its accompanying crook were introduced. The fagging stick, a light wooden rod with the end turned to a right angle, was held in one hand and used to gather together enough corn to form one sheaf, or to bend the uncut corn into a convenient position for a

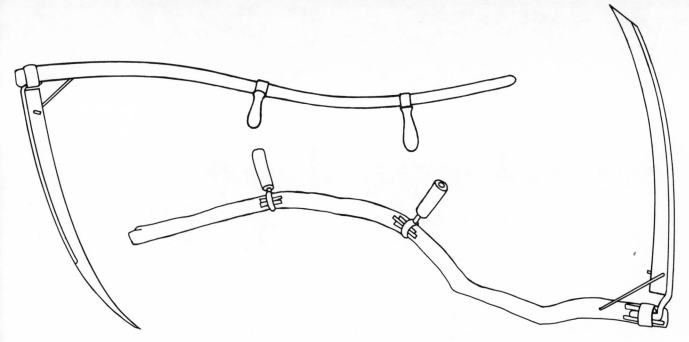

Mowing scythe

Scythe with a natural wood handle, c.1900

clean cut with the hook. This useful implement disappeared when hand instruments were no longer employed for the harvest.

The scythe stone. Before the season's work could begin it was necessary for the scytheman to render his instruments as sharp as possible. In order to achieve a suitable cutting edge he would first employ a block of sandstone about 15 in. in length and of convenient thickness to be grasped in his hand. It was commonly called a scythe stone, being square or round in section but always broad at the centre and tapered towards either end. The manner of sharpening a scythe was first to stand the blade on its point, holding it firmly with one hand whilst using the stone in the other. The stone was placed flat to one side of the blade at the end nearest the snead and by a series of short stroking movements it was pulled across the blade, gradually working down the entire length. Both sides of the blade could be completed except for a few inches nearest the point; these were then rubbed over when the blade was in a horizontal position. The success of this action depended upon the stone being laid flat to each side of the blade; any transgression would result in the edge becoming round or less sharp. Such stones are still essential whenever the old scythes are brought out.

The scythe strickle or strike was used after the stone to give the blade its final edge. This was not a stone but a piece of oak or other hardwood coated with emery, with one end shaped to form a handle. It was about 4 in. in length and 2 in. in

width and the thickness was such as enabled the scytheman to carry the strickle in his belt. Should the edge of the blade become damaged through contact with stones or dulled through prolonged cutting, the strickle was at hand for immediate repair.

Scythemen often made their strickles with a piece of oak about 12 in. long and 2 in. square. It was fashioned into a handle at one end whilst the flat side surfaces were indented with rows and rows of tiny holes, so that tallow, goose grease or bacon fat would adhere. All four sides were then coated with sharp sand. A strickle of this nature was often attached by a ring and a hook to the top of the snead, whilst the scytheman was working, and a supply of grease and sand was kept nearby in case it needed recoating. Strickles disappeared from the harvest field along with sickles and scythes.

The hand hay rake was constructed with a willow or ash wood handle about 8 ft. long, which forked to support a head made of harder wood. The

The daisy rake

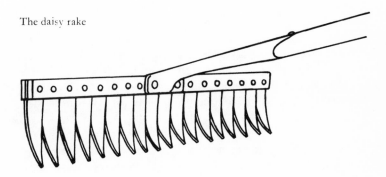

137

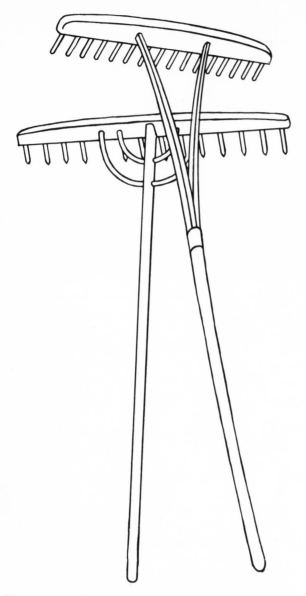

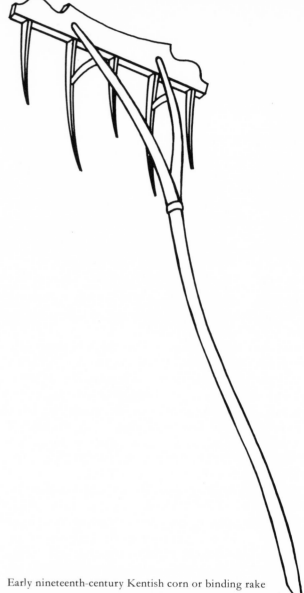

Two patterns of hay rake used in England

Early nineteenth-century Kentish corn or binding rake

wooden teeth were short and had rounded ends in order not to disturb roots and stubble and were screwed into the head to prevent their coming loose or dropping out. In the years when all harvesting was done by manual labour, huge quantities of these rakes were required and their making was then an important rural industry. There were variations of the pattern from county to county with regard to the length of handle (the stail), width of head, the angle of the head, and number of teeth etc. They were employed in the hayfield for centuries before horse-drawn machines were invented to complete the cutting and raking.

The binding rake was also used after the scythe or sickle had done its work and was brought into use when the crop had lain for a sufficient time, and was thoroughly dry. Its handle was shorter than that of the hay rake but its head was wider and the iron teeth were longer, being curved under near to the point. The rake was used by the labourer to collect enough corn to form a sheaf and this he did by taking the rake in both hands and pulling it towards himself. When sufficient had been gathered, he turned the rake over backwards, which placed the corn in a convenient position for tying with a **sheaf-band.** The corn was held securely in the curve

138

of the upturned teeth by the labourer's knee, leaving both of his hands free to tie the band as tightly as possible. This rake was used in southern England, particularly in the county of Kent, during the nineteenth century. Its use does not seem to have extended beyond that time.

The stubble rake was used for gleaning stubble of any cut corn that might have been passed over. Consequently the teeth were set close together and were curved under near to the point, so as not to pierce or disturb the soil. Hand stubble rakes were essential at harvest time until replaced by **horse-drawn rakes** in the early nineteenth century. This hand rake did not however disappear and continued to be made in the traditional form of timber, for which ash was preferred, set with iron teeth. Usually the handle was 6 ft. in length and tenoned into the centre of a head 5 ft. in breadth. The width of this head and the nature of its employment necessitated an iron brace, fixed near to each end of the head and attached to the handle about midway along its length. The teeth were about 7 in. long with curved ends, the better rakes having them screwed through the wood and retained by a nut. Sometimes the stubble rake was equipped with a small wheel on either end of its head in order to make it much lighter and easier to draw across the field. The user pulled the rake along behind him holding the end of the handle in one hand and its cross-piece in the other. They remained essential whilst the harvest was completed by hand and even in years after they gleaned the stubble behind the automatic reaping and binding machine.

The drag rake emerged from the eighteenth century and was similar in appearance to the stubble rake except that its two wooden handles were brought together at the end and braced with a bar to form the shape of the letter A. It was occasionally used to rake over stubble but was intended as a general purpose rake, its robust construction being perpetuated in tubular steel during the late nineteenth century. It was used until recent times for work that did not require a horse-drawn rake.

The pitchfork is still to be found on some farms, but it has been very much limited by the introduction of modern harvesting equipment. Until recent years it was an important item in the field or farmyard, and sometimes called the hay or straw fork. Its long shaft was of considerable assistance for building stacks, filling waggons and for tossing hay during harvest. It had two iron prongs which gathered the straw; their length varied throughout the country, sometimes being broken off short and the ends rounded to ensure safety when working with straw amongst animals. The introduction of the mechanical **hay elevator** made the pitchfork less essential to the farmer.

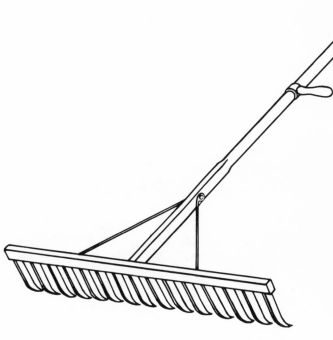

Drag rake

Stubble rake

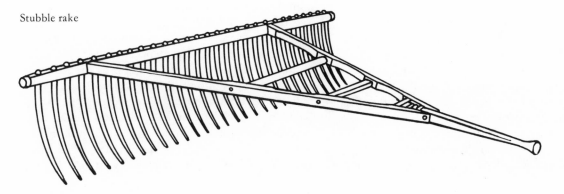

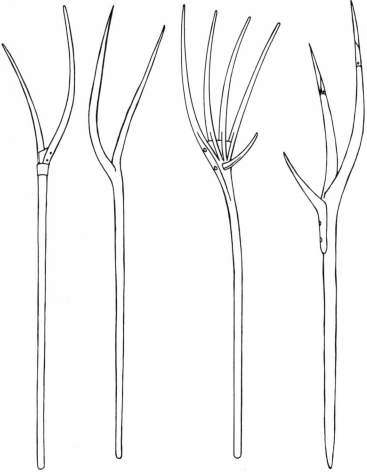

Wooden barley forks grown naturally to shape

The wooden hay fork was more common on the Continent during the last century than in England, where a ready supply of factory-made tools had become available. The wooden hay fork could be easily constructed by any labourer, consisting simply of a sapling about 5 ft. in length and 2 in. in diameter. The sapling was split at one end with each of the divisions being formed into a point. A leather thong was bound tightly around the sapling at the point where the division ended, and a wedge was driven up between the points to separate them. This simple tool was used for working hay and straw.

The barley forks illustrated were nineteenth-century English specimens formed by wood growing to natural shape. A centre prong was included, sometimes on the same plane as those either side, otherwise it protruded outwards to form a catchment as the crop was lifted. The total length of the barley fork was about 5 to 6 ft. Such

forks became obsolete as the harvest was mechanised.

The sheaf gauge was carried by farm supervisors, contractors, and merchants or any persons wishing

A sheaf-gauge

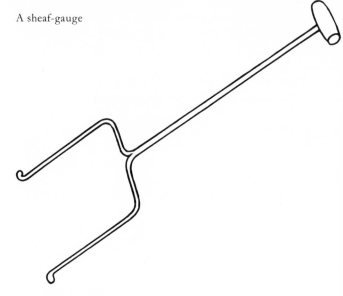

to ascertain that corn sheaves were of the required dimensions, this being 3 ft. in circumference at the point where the sheaf was tied by its band. This instrument was formed in thin iron rod and measured about 3 ft. in length. To form the gauge the iron rod was divided near the bottom into two branches, each branch being 1 ft. in length, and set 1 ft. apart. This division delineated an area 1 ft. square which was open along its bottom side. The gauge was pressed down over the sheaf and was completely filled up if the sheaf was of correct size. This instrument developed from cruder measuring devices and continued in use for as long as corn sheaves were formed.

Sheaf-bands were made from a few stalks of corn and were essential before the automatic binder was invented. They would be made before harvesting commenced by old men or, if the morning was wet and the corn not dry enough to cut, then the reaper and his gatherers would be employed in similar manner and paid a few pence per bundle of 100 bands. It was necessary to produce as long a sheaf-band as possible and they were therefore made from oats or barley pulled up by the roots before the straw was ripe. This would be taken to the barn, where the soil was removed from the root and the grain lightly knocked out with a **flail**. The bands were then made by twisting a few stalks together

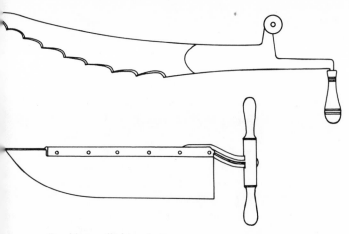

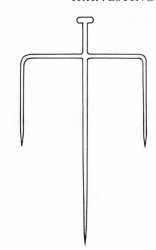

Double-handled hay knives

Hay needle

at the ear-end, with some stalks reversed for added strength, and they were used at harvest time to tie sheaves of corn.

The hay knife. Hay stacks become very compact after only a short period of time and whenever a portion of hay is required for consumption, it is necessary to use a sharp-edged knife in order to remove it. Many patterns of hay knife were available during the nineteenth century, the most popular having a steel blade about 15 in. long set at right angles to a wooden handle. In order to cut effectively, the blade was kept exceedingly keen, and even then a great deal of strength was required to pull it through the solidified hay. Incisions had to be made along all faces before a block could be removed. Stacking and subsequent removal of hay or straw was made much easier by the invention of the automatic baler. Hay knives were retained by those farmers who continued to stack their hay in a loose state.

An American hay knife known as Weymouth's Patent was recommended to British farmers after 1880. It was recognised to be highly efficient for cutting from the stack or bale, and for this reason it had been awarded the first premium at the 1876 International Exhibition in Philadelphia. It was fashioned from one piece of steel about 2 ft. 6 in. in length and the cutting edge was furnished with serrations which became progressively smaller as they neared the pointed tip. Two wooden handles were provided, one for the right hand at the rear end of the blade, whilst the left hand grip was some inches forward and came out from the side of the blade.

Hay needles were usually made to the shape illustrated, though they were often nothing more

than a 3 ft. iron spike with a ring or knob at one end. They were plunged deep into hay and formed a handle by which the truss was lifted, or they were used to secure ropes to a waggon load of hay and various other requirements. Modern bales of hay do not require such fixtures.

The stack trimmer was used by the builder to ensure neatness in the final shape of his corn stack. It was necessary for him to inspect the outer sides of the stack at intervals and to adjust any sheaf that had become disarranged or out of line. This simple operation could be performed by his hand until the height of the stack and resulting pressure towards the base made the task impossible. The stack trimmer, a thin, flat wooden board some 18 in. long and half as wide with a handle, was then used to beat back firmly the butt end of any projecting sheaf and thus produced a smooth finished wall around the outside of the stack. The old method of loose stacking is seldom carried out now. The modern farmer simply piles his bales of straw one upon the other.

The rick borer was of importance to every farmer and was used to ventilate a heated haystack which would have acquired a dangerous amount of combustible gas. The borer replaced a pointed or iron shod staff that the early farmers employed. This was rammed and hammered for almost its entire

A stack trimmer or beater

length down into the stack, and on being withdrawn it left a circular hole in the hay through which the dangerous gases could escape. The rick borer was probably developed during the middle of the eighteenth century when a similar tool was just being exploited for the purpose of **land drainage**. Both were designed to penetrate and withdraw a core of material. The rick borer had a cylindrical shell cutter, about $2\frac{1}{2}$ in. diameter and 18 in. in length, that was attached to an iron shank 1 in. square and was screwed down into the hay by means of a cross-handle. Extension rods were provided, each with a male and female coupling joint, so that the stack could be pierced right through. The cutting shell had to be removed frequently from the shank and the plug of hay extracted by means of a drawing screw, which was lowered and turned in the same manner as the cutter. The finished hole was neat and precise, and several of these drilled about the stack would guard against the risk of fire and the loss of a valuable crop. The borer was useful for cutting holes into large haystacks which had become too solid near the base for the stack thermometer to be inserted by hand.

The stack thermometer was a simple but effective instrument for recording the temperature of a hay or corn stack. It was an iron tube about 7 ft. long and 1 in. in diameter, with a bayonet point at one end and a cross-handle at the other. A thermometer was located inside the tube immediately behind the point and the movements of its quicksilver could be observed through a longsight-hole cut into the tube. The temperature of 170 degrees was marked in red, and above this point the stack was considered dangerous, with ignition taking place near to 200 degrees. The instrument was inserted horizontally into the stack and it was necessary to twist and turn on the cross-handle until it had pierced to the desired distance, whereupon the thermometer would be left for a short time. When it was withdrawn, the temperature would immediately be recorded on paper, a series of which would indicate whether or not the stack was liable to become unsafe. The stack thermometer introduced about 1870, at the cost of 25s., represented an excellent investment for any farmer. Hay is seldom stacked to such density and for such lengthy periods nowadays, so there is much less risk of fire.

The harpoon fork was an instrument by which large quantities of hay or straw could be lifted directly from a wagon and placed in any position on the stack. It was widely used before the mechanical elevator was introduced, and even then was still popular on small farms well into the twentieth century. Formerly, the entire harvest of hay or straw was taken from the wagons and raised aloft by the use of **hand pitchforks**, a tedious task that involved many labourers and consumed a great deal of time. The new invention was of enormous value to all farmers, especially those who managed their farms with little or no assistance, as it enabled one man, together with a horse, to elevate the whole contents of a wagon in a short space of time. Most of the development towards an efficient harpoon fork took place in America, where several patents were taken out during the 1860s. The first patent was issued to E. L. Walker in 1864 for a single harpoon fork which was subsequently called the Nellis fork, and a double harpoon fork came from S. & E. Harris in 1867. After about 1870, British farmers purchased a single harpoon fork made by Messrs Morton and Coleman, based upon Walker's American patent. It was simply a pointed steel bar, about 4 ft. long, 3 in. wide and $\frac{1}{2}$ in. thick with a large ring at the top for attachment to a rope. A small section nearest the point moved on a hinge and through the action of a lever arm could be brought up at right angles. When in use, the farmer would plunge the fork as far down as possible into a load of hay and then pull up the lever arm which caused the pointed section to turn outwards at a right angle. This then provided a support for one load of hay which was lifted from the wagon as a compact lump clinging about the fork. After being hauled aloft by means of a horse-drawn rope and pulley, the load was directed into position ready for its release onto the stack, which the operator effected by pulling a cord attached to the lever arm. This action returned the point to its former position, in line with the main bar, and the hay being then left without a support slipped down from the fork onto the stack. The fork was then lowered down to the wagon, to pull up another load. Double harpoon forks worked in similar manner, but they supported almost twice the quantity of hay.

The farmer was left to provide his own device for elevating the fork and its content. The common practice, in England, was to take a pair of long poles joined together at the top and lean them over the stack. With a pulley attached at the highest point, this arrangement served as a crane and sometimes a swinging jib was included. Various forms of guide-rail track were developed in America during the 1870s. It was permanently secured to the ceiling of a high barn or other suitable shelter, and the harpoon fork, attached to it by small rollers, was easily moved to any part of the fork. Lifting arrangements have remained essential in the barns and stack yards of large farms.

The pea hook. Until mechanical means became available in the twentieth century, a pea hook was used for cutting down a field of peas or beans. Owing to the irregular manner in which these crops were allowed to grow, any other instrument such as a scythe or sickle used upon them would result in considerable damage to the pods. The pea hook was little more than a long, sharp knife, joined by a socket into a wooden handle, curved and broadened at the extremity of the blade. The user would slash off the plants individually some inches above the roots, and the pods would later be

stripped off by gatherers. Green peas are still harvested in the same manner, but hardier varieties for drying and stock feeding are now harvested by machinery.

The caving fork was also known as the cocking or pooking fork. It was similar in height and construction to the common fork except that its three tines were set wider apart. They were made to continue up above the foot tread for several inches before curving over in a half circle to form a scoop. This fork was particularly useful for taking up 'cavings', a name given to short straw left over after threshing.

The pulse fork had four slender iron prongs, each about 2 ft. in length and 5 in. apart, with a wooden handle about 2 ft. 6 in. long. The large head of this fork was useful when moving pulse, peas, and beans etc. which were bulky but very light in weight. The caving and pulse fork became less essential as reaping and threshing were completed in one operation.

The potato graip. Before mechanical means became available in the eighteenth century, the long and tedious work of lifting potatoes from the earth was performed by an army of labourers. For this

Left: caving fork
Right: the pulse fork was light in weight and used for moving pulse, beans, peas, etc.

Potato graip

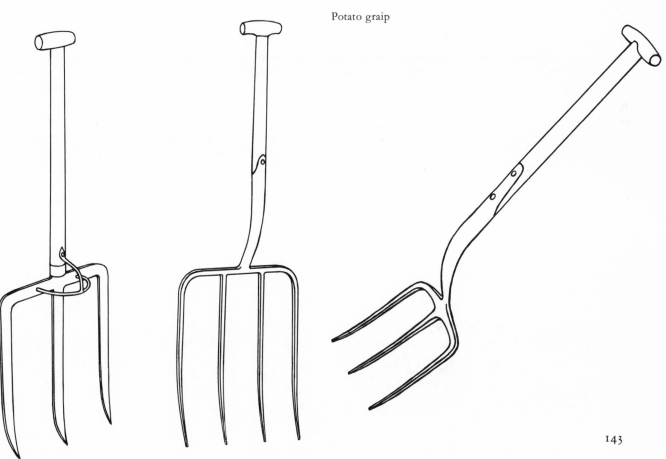

143

purpose, each man was provided with an instrument called a graip. It closely resembled a pronged fork, having a similar helve and a digging head with three prongs. Its prongs however were flatter and made wider towards the ends, before tapering to a point. This flattened form of prong effectively turned the new potatoes out of the earth, which were then gathered into large baskets by women or boys. The baskets were made with coarse woven wicker or rushes to the shape of a half sphere, with a single handle across the open top. The potato graip continued into this present century. No farmer would ever allow his potatoes to be harvested with spades, as a closed blade easily cut through the tubers, whilst the wide set prongs of a fork did not. Prongs also pushed into the ground much more easily and deeply, leaving the turned-up land in a loose desirable condition to stand the winter, or be sown straight away with corn. Potato forks are still available for the market gardener but are of little use to the large scale potato grower.

The iron hand was provided as an aid to the potato gatherer in order to prevent damage to his own hand through prolonged scratching and groping amongst the soil. It was made in iron to the size and shape of a human hand, or claw, with a short thick projection at the wrist which the user grasped in his own hand. It was first used about 1838 with the **Rackheath** plough, but on the whole was not so practical or as widely used as the graip.

The potato grading shovel was somewhat similar to the common shovel, with a wooden helve and

Potato shovel

cross-piece, but its blade was made in the manner of a grill. It was composed of parallel iron rods, with one set crossing the other at right angles, to form square compartments through which small potatoes could pass. This pattern of shovel was used to sort seed potatoes from those intended for market. The illustrated potato shovel had a scoop-like head 12 in. wide and 16 in. long, formed by parallel iron rods. It retained a large number of potatoes but allowed the smaller ones to fall through. This tool was indispensable when filling sacks full of potatoes and was often used in conjunction with a large wickerwork funnel that served to direct all of the potatoes into the mouth of the sack. Grading shovels are still much used.

The peat spades. Peat was an important form of fuel for many centuries, and in some districts of England and Scotland it was preferred to coal, even when the latter could be purchased from small mines nearby. Peat was formerly cut in very large quantities for use in lead smelting, lime burning and various other industries, but during the present century it has formed a cheap domestic fuel for those people, mainly farmers who live in remote areas away from collieries and transport routes. Peat cost the consumer nothing except the time spent extracting it from the ground. A spade was the only essential tool for harvesting this material, and it can be noticed in the illustrations that the shape of blade varied considerably from one district to another. Not all were provided with a foot tread, as peat is soft and can generally be cut by the pressure of arms and shoulders alone, but all blades had a flange or wing at the side so that two sides of the block could be cut with a single stroke. The peat blocks varied in size, but one suitable for an average farm house range was about 12 in. long, 8 in. wide and 2 in. thick. The user first worked along the upper edge of the pit cutting downwards into the peat, then standing on the bottom of the pit he under-cut the blocks with a horizontal action and proceeded to cast them onto a low peat barrow or a sledge placed conveniently near the work. The broad area near the bottom of the spade handle supported half-a-dozen or so blocks, which he carried and placed onto the barrow at the same time. They were then moved away to flat, solid ground nearby, spread out evenly and left for about one week to dry, after which they were hardened sufficiently to be leaned up against each other in twos or threes, so that the wind could blow between

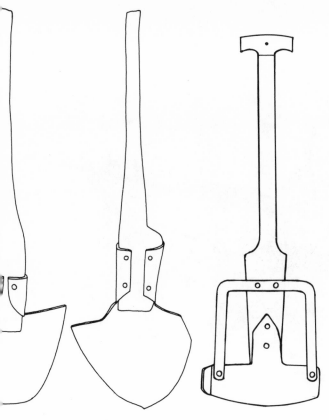

Peat-cutting spades

Hay Harvesting Machinery

Mowing machines or grass cutters were closely bound up with corn reapers during the early years of development. The first machine specifically designed to cut grass was patented, as early as 1812, by Peter Gaillard of Lancaster, Pennsylvania, who tried to reproduce the action of a scythe. A following attempt was made by Jeremiah Bailey of Chester County in 1822. His was a combination reaper-mower and utilised a sharp-edged disc rotating horizontally a few inches above the ground. Another American, William Manning, came close to success in 1831 with a reciprocating cutter which moved from side to side between guard fingers. This was the principle of Obed Hussey's American mower, which marked the commencement of grass cutting machinery as a separate item in the middle of the nineteenth century.

The illustration is a machine by A. C. Bamlett, typical of most other horse-drawn mowers. It consists of three main parts:

1. The cutting apparatus which contained fifteen fingers to divide the grass into bunches prior to being cut by reciprocating knife edge. The cutter travelled close to the ground and was generally supported by a wheel at either end.

2. The framework with draught pole to the horses and seat for the driver.

3. The gearing contained within the framework by which motive power was transferred from two travelling wheels to the cutting apparatus. The cut grass was laid in swathes and later turned and tedded before being stacked. The first tractor-drawn mowers were similar in design and trailed behind the haulage power. Present-day mowers, which still use the clipper principle, are mounted beneath the tractor body slightly in front of its rear wheels, whilst rotary mowers are attached to a power take-off at the rear of a tractor.

The hay sweep was a contrivance for bringing scattered hay together at the side of a rick or cart. Its simplest form was a rope, or a single curved beam dragged forward by ropes attached to either end. It was not a common implement when depicted by Arthur Young in his *Tour Through the North of England*, 1771. The one he saw consisted of two slightly curved bars, held apart by a series of short vertical bars. It was drawn along between two horses for the purpose of collecting together

them. These small heaps were afterwards brought together to form larger stacks about 4 ft. in diameter, 6 ft. high, and tapered towards the top. Small ventilation spaces were left between the blocks in the stack so that they did not spoil before the peat was carted home much later in the same year. There it was broken down into small pieces for burning or perhaps mixed in with similar-sized coals. The tools and harvesting methods described are still retained in some remote districts of Great Britain and Ireland.

145

Bamlett's No. 4 mowing machine

all the dry hay previously cut and hand raked into windrows. It remained one of the simplest implements used on the farm, but none the less it considerably eased the labour of hay harvest until recently, by which time its use was widespread. The sweep delivered its load to a waggon or to the side of a stack if it were to be erected in the same field. However, this handy machine had one intrinsic disadvantage. Since the sweep was always made wide and without wheels, it could not pass through gateways or travel over rough ground. If the hay was to be tansported to a barn or to a stack outside the field then a **hay sledge** had to be brought into use instead.

Of the many patterns of hay sweep used in England and Scotland the wing sweep seems to have been most common. Its method of operation was similar to that of most other sweeps, in that it travelled along the windrows collecting up large amounts of hay which were duly delivered to the stack. Being light in weight the wing sweep was particularly suited to hillside work and uneven ground. It consisted of a timber framework, some 6 to 8 ft. wide with extensions or wings 4 to 5 ft. long, loosely affixed by hinges at either end. The horses were attached by draught chains to the ends of the wings and one horse walked down each side of the windrow whilst the sweep between them collected the hay. When the wings were open wide and sufficient had been loaded, the horses were directed over to the stack and upon arrival the driver took his weight off the sweep and allowed it to ride up over the load of hay so as to leave it all behind in one place, whilst the sweep continued on to collect another load. Thus no time was lost on emptying the hay, as it fell free of its own accord. Some different patterns used until recently, in different districts, are illustrated. Modern processes have rendered them obsolete.

The American hay rake was received into England and Scotland about 1850, though it would appear that a similar implement had been manu-

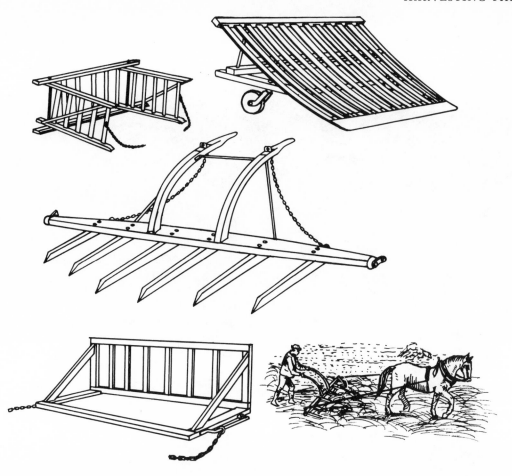

Haysweeps

factured and used in England some years before-hand. The imported hay rake was made entirely in wood, being some 15 ft. in breadth and resembling a gigantic double-sided comb, drawn along trans-versally with the teeth horizontal to the ground. One set of teeth or tines pointed to the fore and collected or swept up hay from the surface of a field, whilst the other set pointed to the rear and remained idle. Usually about eighteen tines, each about 4 ft. in length and pointed at both ends, were spaced along a main beam so that an equal portion projected both to the fore and the rear. The rake was drawn by a single horse, harnessed by chains to the main beam, with the driver walking along at the rear holding the reins and also the two wooden handles of the rake. It was necessary for the driver to apply some weight to the handles, as they were arranged to prevent the rake from tipping over if the front tines should happen to dig into the soil. When sufficient hay had been collected on the front tines, the load could be released by the driver

pulling the handles slightly apart, which freed the backward pointing tines and allowed the rake to rotate forwards. This action left the contents upon the ground and placed the second set of tines in a position to collect up hay. The American hay rake continued well into this present century, having undergone some improvement with regard to the mechanism for releasing the load, but it had no place amongst the growing number of machines in the harvest field.

The hay collector. Whilst the American hay rake grew in popularity, some farmers, especially in Scotland, preferred to construct their own, or purchase a simple wooden version known as the collector.

The hay collector closely resembled the hay rake, except that it contained one row of tines instead of two. There were normally six tines, each about 3 ft. in length, spaced along a beam at intervals of 18 in. Two handles extended back from the beam so that it could be manipulated by a driver as it gathered up

147

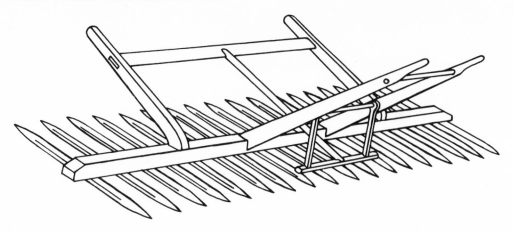

American hay rake

almost half a cart-load of hay. It was drawn by one horse, which was attached by draught chains and swivel links to each end of the beam. The draught chains also served to support the hay, whilst the swivel links facilitated the unloading, which was achieved by allowing the points of the tines to catch the ground and cause the collector to tipple forwards for one complete revolution, during which it released the load. Upon its correcting itself, the driver took hold of the handles and proceeded to collect another load.

Both the hay collector and the hay sweep, which was later made large enough to contain a full cart-load of hay and was fitted with mechanism for lifting the tines, continued well into the twentieth century.

The hay sledge was pulled by one horse along windrows, and hay was loaded onto it for transportation to barn or stack. It was a flat timber arrangement, usually about 8 to 10 ft. long and 6 ft. wide, generally made by the local carpenter. The two runners upon which it travelled were slightly bowed and joined together by cross members, set a few inches apart, as a platform for the hay. The procedure of work required one person, usually a boy, to lead the horses and two men to load the hay. The sledge was directed along between two windrows, whilst the men, one walking on either side, pushed their rakes along the windrow until a 'combing' had gathered together before its head. It was then lifted between arm and rake and placed on two corners of the sledge, the second 'combing' being placed at the other corners and the centre filled in last of all. Three or four more layers were built above in precisely the same way, and the whole

load was secured with a rope thrown over it and lashed to both fore and rear cross-bars. This retained the hay, especially when going up or downhill, until it was untied and the load tipped off sideways at the side of the stack. Sledges were used until recently in many places, especially where there were hills. In many districts it was replaced by the hay bogey, a similar low platform, but one carried at the centre upon two small, wide wheels. Both have been replaced by motorised transport.

Stanging. Hay was carried down from the hills in Derbyshire by a method known as stanging, stang being the colloquial for pole. Two poles, each about 8 ft. in length, were placed on the ground some 3 ft. apart, and a large cock of hay was set across the middle of them. Two men then took up the ends of the poles and carried it like a sedan chair. This arrangement was continued on very steep hillsides and in small fields where the hay sweep and the sledge were impractical.

The horse rake was once a crude class of implement, as can be observed in the drawing of Ketcher's stubble rake. This particular rake represents the transition from hand to horse-drawn rake at the commencement of the nineteenth century, when Mr Ketcher set the pattern for such elegant models as the self-lift rake made about one hundred years later by Blackstone's of Stamford. A distinctive feature of the latter rake was its simple leverage, worked by the driver's hand or foot, in order to raise the tine bar and release the accumulated load of hay or other crop. The steel tines could be adjusted to skim slightly or closely rake the ground and would adjust themselves to irregularities of the surface. This pattern of rake prevailed until tractors

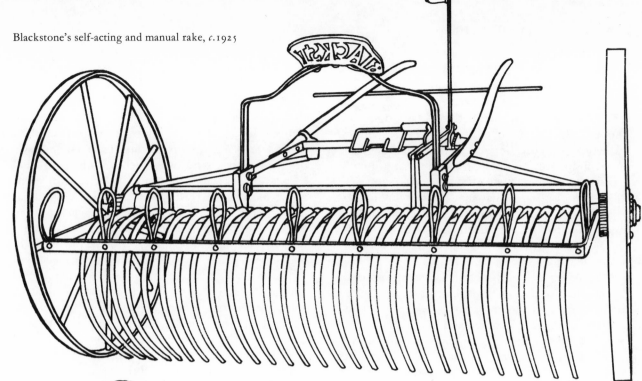

Blackstone's self-acting and manual rake, *c.*1925

Ketcher's bean stubble rake, *c.*1880

were introduced; even then it remained much the same, except that the width was increased from about 8 to 12 ft., and the tines were raised by a trip-cord from the tractor seat. Its wheels remained as before, about 52 in. in diameter. Such rakes are still occasionally used at the rear of a tractor to clean a field of stubble.

The hay tedder was pioneered by Robert Salmon, a Woburn gentleman, during the early nineteenth century, and with it came a new pattern of hay-making. Formerly the grass was cut with scythes and dried out by a long procedure of spreading and gathering, during which it was frequently turned over or tedded by women equipped with wooden hay rakes. The process was exacting and very much dependent on good weather.

Robert Salmon's invention was to reduce the time and amount of labour spent in making hay, and so he designed his machine to follow the scythe, during which it lifted the cut grass from the ground and turned it over so that the side, previously lying underneath, became exposed to the effects of sunlight and more especially breeze. Despite some faults it was observed to spread the hay lighter and more evenly over the ground than could be achieved by hand, and also divided the swath with great rapidity even when the weather was damp. It could turn over thirty acres of grass per day, equivalent to the work of fifteen women, and performed best with the horse going at a gentle trot. Salmon's machine, upon which all later tedders were based, consisted of two large travelling wheels placed at either end of an axle. The axle also contained two smaller wheels placed inside the travelling wheels, and being some 6 ft. apart they were connected together by strips of wood attached to the circumference, thus forming a cage or cylinder around the wheel axle. Each strip of wood was furnished with several iron spikes so that when the machine was drawn along and the cylinder revolved around the wheel axle, the spikes caught up the hay and tossed it into the air.

The tedder became popular in the London area after 1830, but its introduction elsewhere was much slower because many farmers were annoyed by the violent manner in which it scattered hay broadcast, causing it to be blown away by the wind, or alternately smothered both horse and implement. Therefore it could not go well in the wind nor was it calculated to work an uneven ground, as the rotating spikes were defeated by the ridges and lumpy soil, and were also inclined to snap off whenever they encountered a stone or other hard obstruction. Some critics pointed out that it did not necessarily save as much labour as was supposed, for it did not place the hay into rows or any regular arrangement and consequently required twice as much effort to rake it together again afterwards; also in dry weather the valuable seeds were shaken out by its rough treatment.

A Mr Wedlake of Hornchurch, Essex, improved the hay tedder during the 1840s when he divided the cylinder into two sections, each revolving independently of the other and both attached to the same axle by spring mountings, which allowed the spikes to give way whenever they met with mole hills, ridges or rocks. Otherwise it retained similar faults to Salmon's machine, and again hay was blown away by the wind, requiring a great deal of time to be spent on hand raking it together again, a problem eventually overcome by the introduction of efficient **horse-drawn rakes**. Development of haymaking machinery was accelerated in England and America after the invention of **mowing machines,** and some new manufacturing firms were created especially to meet the new demand. Competition between the old and new

brought further improvements such as Smith and Ashby's tedder, in which the revolving cylinders were thrown out of gear whenever it was turned or reversed. Ashby, Jefferie and Lake enveloped the top of their machine with a canvas hood, which caused the hay to fall away evenly at its rear instead of being cast in all directions. Nicholson's double-action haymaking machine was renowned for its even balance and smooth action, brought about by replacing the spiked cylinders with radial forks. This machine gained a prize at the 1857 Salisbury Show, and similar forks were subsequently used by most other manufacturers. The canvas hood was soon replaced by one in lightweight metal and the wheel gearing became more complex. The addition of ratchet escapement on the radial forks allowed them to continue turning after the machine had halted and ensured that all hay was cleared away from the forks instead of falling and twisting about the axle. If the weather was fine the tedder followed close behind the mowing machine, strewing the cut grass quite evenly. One or two turns lifted the grass ready for the horse rake, the implements between them doing the work of a dozen or so hands. Tedders were purchased as a separate implement until the 1930s, by which time other efficient harvesting machinery, the **swath turner** and **side delivery rake,** had appeared in the field. The latter implements have remained essential in the mechanised hay field, though their design and performance have been much improved in late years.

Blackstone's 1920 combined swath turner, collector and side delivery rake

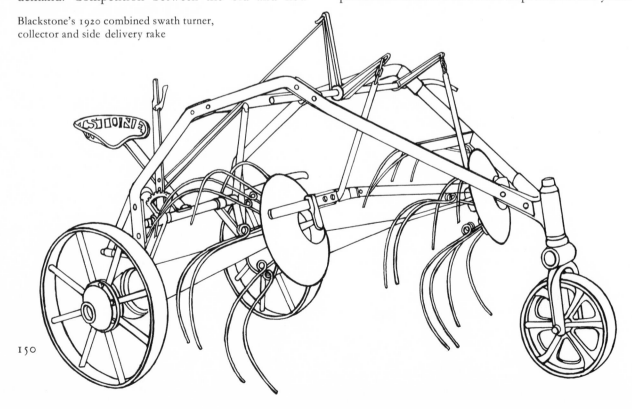

The hay kicker probably originated in America during the 1860s but did not achieve popularity in England until the last decade of the century. It was not so violent in its action as the established haymaker, and as can be seen from the illustration it was a light machine, well within the power of one horse. The series of spring tine forks at the rear were motivated by gearing and crankshafts from the travelling wheels and 'kicked out', throwing the hay up behind the machine rather than over it. This implement was superseded by the **swath turner**.

The swath turner was developed in the early years of the present century to invert the heavy swath left by the mowing machine. If the weather was favourable, the swath was turned over on the day after mowing and turned again for a second time on the third day, but if the atmosphere was damp, work was delayed with a risk of losing the crop through decay. The first form of swath turner had a simple iron framework, carried upon two travelling wheels about 3 ft. in diameter and 8 ft. apart, with a driver's seat high above the centre of the wheel axle. The two rakes, which revolved in a clockwise direction and turned the swath over butt-end first, were situated side by side at the rear end of the framework. The distance between them was adjusted to suit the width of two swaths, and they were activated by means of gearing and horizontal drive shafts from the travelling wheel axle. Each rake consisted of four arms, radiating from the drive shaft, each arm terminating with six steel tines or prongs, set to turn over the swath without catching the ground. This implement is extremely important to modern farmers and has been very well adapted to tractor haulage.

The American Keystone hay loader was a pioneer implement produced in 1875. A large number were made and used in America, and various manufacturers in Europe made them under licence before producing their own versions. It represents the general pattern of most succeeding loaders except for differing mechanisms that caught up the hay, and the subsequent development of telescopic or expanding hay loaders, which could be adjusted for height. The Keystone was furnished with a single centre drawbar that hitched onto the rear of a horse-drawn wagon so that both proceeded along the windrows at the same speed, and whilst some weight was supported by the wagon the loader was equipped with its own two wheels. They were

about 3 ft. in diameter and between the wheels, attached to the same axle, was a cylindrical cage-like structure. This was composed of iron hoops, 2 ft. in diameter, joined together by iron bars which were each set with ten pointed iron tines like those of a pitchfork, to catch up the loose hay in the windrows and lift it as the cage revolved. The wheel axle also carried the supports for the hay elevator, which was some 15 ft. in height and which extended upwards over the rear end of the hay wagon. The framework of the elevator was constructed in timber, with two main beams at either side extending from the axle, where they were secured between the wheels and the cylinder, to the topmost point of the elevator where they were joined together by a roller. On either end of the roller and inside the main beams there was a toothed wheel, each one carrying an endless chain which passed down the full extent of the elevator framework to pass around the circumference of the cylinder on the wheel axle. There the links were caught up by hooks around the circumference of the cylinder and the chains, which were connected together at intervals by wooden slats, were forced to travel to the top of the loader, where they passed around the roller, and returned as the wheels and lifting cylinder revolved. Before the revolution of the lifting cylinder was complete, the hay caught up on its tines was deposited on the moving slats, and being prevented from falling through by the presence of joining wires, it was conveyed upwards to the point where the chains and slats passed around the roller and from there it was tipped down into the wagon.

The Deere hay loader, manufactured by the Deere and Mansur Company of Moline, Illinois, was introduced about 1893 and employed a different principle from that of the Keystone. It, too, was attached to the rear of a moving wagon and the motion of its two wheels was transmitted through toothed wheels and chains to a crank-shaft which was mounted across the full width of the implement. This crankshaft was furnished with eight rakes and its angular shape and rotary motion caused the rakes to move backwards and forwards over the tail of the loader, four of them extending out to gather up hay from the windrow and draw it onto the frame, whilst the four alternate rakes were fully retracted and were placing their collection on the moving web of the elevator. The elevator web obtained its movement from the reciprocating

THE DEERE LOADER IN OPERATION.

THE DEERE HAY LOADER.

THE BECK HAY LOADER.

THE SANDWICH "CLEAN SWEEP" HAY LOADER

THE KEYSTONE HAY LOADER.

Hayloaders. From R. Ardrey's *American Agricultural Implements*

action of the eight rakes, each of which was furnished with hooks on its underside, and as they were in the process of drawing in the hay the hooks located into slots on the elevator web gradually moving it aloft until the hay tumbled over the end and into the wagon below. Hay and straw loaders were replaced, after World War II, by automatic balers.

The hay baling press. The development of mechanical devices to compact loose hay into bales was centred in America, where some patents were taken out during the first half of the nineteenth century, but it would seem there was no large call for such a machine until after 1850. When the demand finally came it was through the necessity of delivering hugh amounts of hay from the rural areas into the rapidly developing cities where a considerable number of horses were used in transport and industry. Although the experimental work achieved by early inventors had proved the baling press to be practical, manufacturers on the whole seemed slow to realise the potential of the expanding haymarket, and a really efficient machine did not emerge until 1872.

H. L. Emery of Albany, New York, was perhaps the first manufacturer successfully to produce a hay baling press and commenced production in 1853. This was a horizontal press in which pistons at either end of a chamber were slowly brought towards the centre, compressing the hay into 250 lb. bales. The piston movement was obtained through a system of chains and pulleys connected to a **horse work**. It could produce five bales in one hour and required regular attention of several men. A continuous form of hay baler was brought into practical form by P. K. Dederick in 1872, and his was immediately accepted as the premier machine in America, whilst its reception in England was equally as enthusiastic. A vertical press worked by horse power was produced in 1866 by George Ertel of Quincy, Illinois, who subsequently devoted much of his attention to the improvement of machines in this field. Steam driven 'perpetual' presses were introduced about 1884 in response to the need for faster baling along with increased capacity. When the bale was sufficiently formed it was automatically ejected at the rear and hand tied with wire. The most successful maker for the remainder of that century was the Famous Manufacturing Co. of Chicago, who, by 1894, had

Savage's straw elevator

153

more than twenty different types of baling press on the market, mostly worked by steam power, only a few by horses. This essentially American development was made portable early in this present century and is nowadays a common sight working at the rear of a tractor.

Straw elevators were constructed on the conveyor principle for raising sheaves or loose straw from ground level to the highest part of a stack. They developed during the nineteenth century and accompanied the threshing machine from one farm to another. Modern elevators for raising bales and sacks etc., are much the same as the one illustrated except that some are telescopic and driven by a petrol engine.

Root Harvesting Machinery

The potato digger. The laborious task of digging potatoes with a hand graip was assisted to some extent by using a plough to break the ridges open first. Later developments, such as the **plough graip** and the brander, which could be attached to a plough for turning the potatoes out of the soil, were a considerable stride forward. The next achievement was a machine patented in 1852 which employed an elevator to remove potatoes from the soil. This was followed three years later by a rotary digger patented by a Mr Hanson. His machine was remarkably similar in its working action to the modern potato spinner and it would seem that this machine set the pattern for future development, whilst the elevator principle of the former invention came to be used in the **potato planter**. Mr Hanson's digger was carried at the front end, on two small iron wheels, and two larger wheels were placed midway along its length. The spinner was at the rear of the machine and received its rotary motion via a system of gears and a long horizontal drive-shaft from the two large wheels. The spinner was formed by eight forks which radiated from a disc at the end of the drive-shaft, and was positioned so that the points of the forks were able to dig about eight or nine inches below the surface of a potato drill. The precise depth of penetration by the spinning forks could be varied by raising or lowering the front wheels, which affected the balance of the machine upon the axle of the two large wheels and so raised or lowered the rear end where the spinner was located. An iron share set at the same depth preceded the spinner as the machine was hauled forward, serving to break down the soil and loosen the potatoes in the drill. They were then sorted from the soil by the spinning forks and thrown sideways, where they struck against a net hanging at the side of the machine, and this caused them to fall in a tidy row alongside the opened drill. Farmers at this time were alarmed by the violent manner in which this spinner threw the potatoes, but it is probable that this machine damaged the crop no more than a plough. Despite criticism the rotary spinner was exploited and the potato plough was gradually cast aside, although it was used to some extent up to the early part of this present century.

During the remainder of the nineteenth century,

horse-drawn potato diggers were made and used in large numbers. They became more compact and lighter in weight, whilst the wheels, framework, and mechanism were made entirely in iron or steel. Its operative parts consisted of a 'share' which penetrated below the depth of the potato roots and raised them up, whereupon they were separated from the soil by the revolving tines at the rear of the machine. But despite the improvements in construction and draught, a disadvantage existed. This was the inability of the spinning tines to reach the potatoes lying at the sides of the rows and consequently they remained uncovered after the passage of the machine, unless the share and digging tines were set to an impractical depth. This fault remained until the early years of the twentieth century, when a new arrangement for the tines was discovered and found to be highly effective. Until that time the spinner had consisted of a number of radial tines attached to the end of a horizontal shaft, each being identical in length and angle so that their ends scribed a circle when the shaft was rotated. The improved form was called 'link motion' and allowed the tines to follow an eliptical path, which enabled them to dig over the full width of the drill and effectively remove all potatoes from the centre and the outside of the rows. Manufacturers exploited this new principle of 'link motion' and some very effective machines were produced, to be worked by horse and later by tractor. During the 1920s two old principles were revived. Some manufacturers reverted to the idea of a spinner with radial tines, but tilted backwards, in the frame, to achieve an eliptical rather than circular digging

movement. The elevator type of digger was also revived wherein the soil and potatoes were raised up by mechanical shovels and separated through a series of sieves, the soil falling back through the meshes whilst the potatoes were laid on the ground at the rear of the machine.

The Bamlett royal prize potato digger, as produced in the early years of the present century, was a very popular machine and worked on the new link motion principle. It consisted of a steel framework with a small swivel wheel at the rear and two large wheels at the front end, with a lever-operated arrangement to raise and lower the spinner and the share. The spinner was comprised of two discs, one attached by its centre point to the end of a drive shaft, whilst the second, with the digging tines attached, was held about 6 in. apart and joined to the first by five link arms, through which it obtained its eliptical motion. The rotary speed could be adjusted by simply changing a sprocket wheel which would ensure a digging speed consistent with the natural pace of the horse. A gear too high would have damaged the potatoes and spread them too far for quick collection, whilst a gear too low would not unearth the potatoes sufficiently, leaving the land in ridges with many potatoes still buried. Although the digging action of the 'Royal' and similar machines proved most thorough, some difficulty did arise with the large number of moving parts that were present, and due to the constant barrage of soil and subsequent entry of particles into the bearings, the machines were quickly damaged. A 1920s' potato spinner is illustrated. They have changed little in design since then.

Blackstone's 1920 combined swath turner, collector and side delivery rake

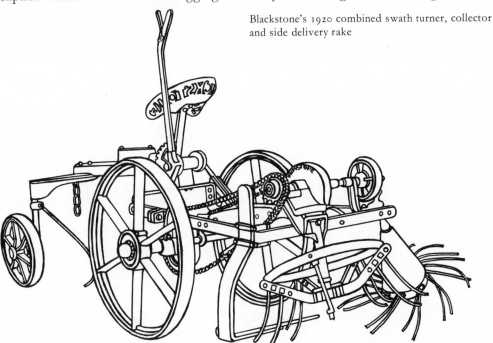

155

Stacking Hay and Corn

Most farmers prided themselves on being able to construct a high stack of sheaves that could stand of its own accord. Supports were occasionally employed at the centre of a rick and sometimes took the simple form of a wooden post, held in a vertical position by means of its base being let into the ground. The stack was then built around this crude pillar. This method was not widespread and was practised only by the worst farmers; the majority endeavoured to set their skill in constructing a stack without support, leaving the sheaves in such a perfect manner as would ensure they were well ventilated. Rick supports of a kind were employed as an aid to ventilation of a stack. These were usually made by a farmer in his stackyard. His aim was to create a hollow area at the heart of the stack which would draw in air from the exterior and thus keep the crop in good condition. A common form of support was made with three posts or tree trunks, each about 7 ft. long, cleanly cut off at the top and bottom and stripped of all branches. The three trees were stood upright with their butt-ends about 3 ft. apart from each other and were then leaned inwards until the tops were touching. At this stage they completed the outline of a pyramid, which was firmly retained by lashing the tops together with rope and nailing planks of wood horizontally at intervals around the outside. These planks also served to prevent any sheaves from falling into the support when placed upon a **stathel** and the rick built around and over it. The sharp point that the pyramid presented was, however, liable to pierce through the top of the stack as the hay consolidated.

Building round stacks. Not all stacks were built with a support at the centre, and most farmers could pride themselves on being able to construct a stack that would stand of its own accord. Their method of construction was to set down a sheaf on its butt-end in the centre of a stathel, placing more sheaves around it which were also upright, but inclined very slightly towards the centre sheaf. To complete the circumference of the stack further layers of sheaves were added, each with its butt-end lying towards the outside of the stack and the ear end bent up against the centre sheaves. The builder was required to compress the outer layers of sheaves to a greater density than those in the centre, and thus provide a protective wall against birds and the bitterest of elements. He would then continue to build up a number of layers in a similar manner to the height of the eaves. When this point was reached the layers were progressively overlapped towards the apex, with the butt-ends still to the outside, until the top of the stack was covered over. After the final crowning sheaves had been put into place the stack was thatched over and tied with straw rope. It then remained untouched until required for threshing.

Some farmers preferred a hollow at the centre of the stack, and they achieved this by laying the sheaves up against a bale of straw rope which was gradually drawn up through the centre of the stack as the layers of sheaves were applied. When the bale of rope was finally removed, it left a hole about 2 ft. in diameter from the bottom to the top of the stack, through which heated air could escape and so allow cooler air to flow in from beneath the stathel. This slow but picturesque method of stacking is not now necessary with the use of hay and straw balers.

The rick stand or stathel. When stacking corn, farm workers made great effort to produce a regular, tidy rick. The achievement was considerably reduced and the crop damaged if dampness or vermin attacked the rick at its base. In order to avoid this danger, the rick was built upon a stand or stathel which stood about 2 ft. above ground level. Before cast-iron rick stands became available the early farmer out of necessity improvised his own version, using timber and stone. It was essential to select flat, firm ground before building could begin, as the weight of the construction and the rick upon it was considerable. Firstly, eight flat stones were positioned in a circle to the diameter of the intended rick and a ninth stone of similar character was positioned in the exact centre of this circle. Then a pillar of stone 18 in. high and 1 ft. square was placed on each of these foundations, followed by a flat capping stone, 2 in. thick and 18 in. in diameter on top of each pillar. The timber framework of the stathel was then put into position. This was achieved by arranging heavy beams of timber in a radial fashion from the central pillar. The inner end of each beam thus rested upon the central pillar with its outer end on one of the pillars that comprised the circle. A floor was then laid upon these beams by using stout planks which were set about 6 in. apart and held with nails. This type of stathel was intended as a permanent structure and was seldom moved once it was built. If treated with an oil or preservative, it would last for many years before the timbers had to be replaced. Whole granaries were often raised on stathel stones.

Cast-iron rick stands became available early in the nineteenth century and were made in many intricate formations to suit the building of round and oblong stacks. They retained the main principles of the early stathel, but had an iron-work platform raised on stout iron pillars at intervals around the perimeter. The pillars rested upon flat base plates and were elegantly tapered towards the top, where a large capping dome prevented the most agile of vermin from passing into the rick. An advantage of this type of stand was that it could be shortened as the hay or corn was removed. This was achieved by simply lifting away each section of the stand when it became empty, which allowed a wagon to move in close against the open side of the rick.

A cast iron rick stand

Farmers in Norfolk continued to use a cheap but effective method of securing their corn ricks against rats and mice long after cast-iron stathels were introduced. Their method had been used for generations but it gradually declined in that area, as improved methods became known. The traditional expedient required a rick to be built with vertical sides. It was afterwards cut all the way round so that the lowest 2 ft. tapered inwards for about 18 in. at ground level. The area where the stack had been cut away was daubed over with clay or mortar, until it formed a layer about 2 in. thick, with a rim, or projecting edge, at the top. When it had dried, the whole of the wall was whitewashed over and formed an effective barrier against all vermin.

When iron rick stands were first introduced, manufacturers tried to promote the idea of fitting them with wheels. They succeeded to a degree and some farmers went to the expense of laying iron rails, similar to railway tracks, between their stacking area and threshing machine. The iron stand and the corn rick they supported were moved, by means of four horses, to the shelter of the threshing barn, where the rick could be safely dismantled without fear of wet weather. However this advantage and that of time saved in carting hardly compared with the cost of laying tracks and maintaining them in working condition.

Mr Crosskill, the implement manufacturer of Beverley, promoted a type of iron railway designed to be laid down and then taken up with ease. He intended it for bringing crops to the farmyard, but it did not catch on to any great extent. The price of this railway set in 1853 was £54.10s.0d. and included 150 yards or 30 lengths of rail at 4s. per yard, one horse-drawn truck to tip either side, one horse-drawn truck to tip either end, one turn-table, extra points and branch lines etc.

The stacking stage. Whenever hay or corn stacks were built, stacking stages were found to be of more use than ladders as they rendered the work a deal easier and also allowed two or more men to work side by side in an elevated position. Stages were made by the farmer himself and consisted of two timber beams a few feet higher than the intended stack, and a movable platform or stage which was

Stacking stage

157

attached to the beams by hooks and chains. The beams were held apart at a distance of 5 or 6 ft. by iron cross-straps and inclined at an angle towards the apex of the stack, as shown in the illustration. The staging was secured by the use of two iron pins which located into holes in both beams, and it was further supported by hooks and chains on either side. As the height of the stack was increased, so the platform was raised by locating pins into other holes nearer the top of the beams.

Rick clothes were kept close at hand when stack building was in progress, since it was essential to provide some protection against inclement weather. In order to cover an unfinished stack at short notice the farmer would provide two tall poles, each set by its butt-end into the centre of a heavy waggon wheel. The poles were positioned in an upright manner at either end of the stack and required no other attachment but three guy-ropes which were fixed to the top of the poles by iron bands, with the other end pegged down to the ground. Each pole was equipped at the top with a pulley wheel, over which ran a long length of rope. When rainy weather was imminent, a third pole over which was hung a huge waterproof canvas could be raised above the stack by means of the ropes and pulleys. It was then opened out to form a tent-like cover, and was held secure by ropes in that position until the weather had cleared and stacking was again possible. Bales of hay are now stacked beneath the cover of a closed or open-sided barn.

Covered shelters for hay and corn were promoted by various British manufacturers during the second half of the nineteenth century. They declared stack thatching to be wasteful of straw and unsafe in such a variable climate, so for that reason some farmers purchased a 'Dutch barn' or open-sided shed. The pillars and framework were of wood or iron with the roof covered in the new corrugated iron sheeting that was rapidly superseding the use of slates, tiles and boarding, in farm building. Modern farm buildings are made in steel and concrete.

6

Processing the crops

Nineteenth-century threshing scene. From *The Popular Encyclopedia*

Threshing and Winnowing

Threshing is the operation or process by which the seed is separated from the stalk.

The earliest known method of threshing was to beat the grain from the ears of corn with a stick. The Egyptians improved upon this by spreading out the loosened sheaves on a wide area of firm ground and then driving oxen over it to tread out the grain, a practice which must have inevitably damaged or dirtied a large proportion of the crop.

The next development was a kind of sledge which the oxen dragged behind them over the corn. Flints were securely embedded in its underside and the driver added his weight by standing on the rear edge.

The Romans used their *Tribulum*, a heavy wooden platform mounted upon rollers, which was also dragged by oxen over the piles of sheaves. Threshing boards were similar constructions and formed a common feature in most Mediterranean countries up to this present century, when they were generally displaced by modern implements and techniques.

The flail was an instrument which threshed corn more effectively than a straight stick or the feet of tramping oxen. It most likely developed in Europe during the Middle Ages and continued to be used extensively up to the late nineteenth century. The gradual perfection of the mechanical threshing machine during that century left only a few small farmers in need of the flail, though it was retained by many for such occasions when the use of mechanical means was impractical. The flail was of simple construction, having a hand staff of hard wood, usually ash, 4 to 5 ft. in length and which the labourer gripped in both hands. The swiple or beater, the part which struck the corn, was a rod 3 ft. in length and $1\frac{1}{2}$ in. in diameter, for which thorn was considered most suitable. The hand staff and swiple were bound by caplins or leather thongs, and these in turn were tied together by a middle band of eelskin thongs in the form of a loose or flexible joint. An alternative method of attaching the two components was to mount one end of the swiple with an iron strap, fixed to the wood with nails and projecting to form a loop through which the eelskin thong was passed. The reason for this flexible joint was to enable the user to swing the beater around his head before bringing it down

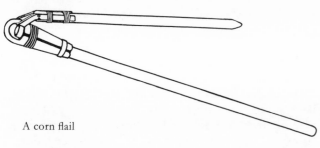

A corn flail

with crashing force upon the corn in order to knock the grain from the ear. Threshing with a hand flail, skilled but laborious work, was carried out in a barn with a specially prepared floor and wide doors at either end. A constant flow of air between the open doors provided a natural winnowing process by blowing the chaff away from the grain. Flails were generally replaced by mechanical threshers before the end of the last century.

The threshing floor is often mentioned in the Bible. Gideon's threshing floor upon which he laid the miraculous fleece of wool, *Judges* 6:39, is a good example. Those used by the Egyptians, Hebrews and the Romans were not covered over, only surrounded by a low stone wall to keep animals away. They were about 30 to 40 yards in diameter, sloped slightly from the centre to the circumference, and were made with tightly packed earth at the summit of a hill. Specially constructed threshing floors were employed in most countries up to this present century and were an important feature in any village, commune or farm where mechanical threshers were not available. In many cases they were constructed inside a barn midway between two large doors, the size of floor being proportionate to the amount of flails expected at work, but always the surface was smooth and perfectly firm. Some new materials were tried: flat rock, mortary compounds, brick and stone were found to bruise the grain and none of them encouraged the flail to rebound. The older method was to excavate an area 6 to 9 in. deep and fill it half-full with dry sand or gravel, on top of which a mixture of clay, sand, chalk, gypsum, cow-dung, chaff and bullocks' blood was applied in layers $\frac{1}{2}$ in. thick. Each layer was rammed down and allowed to dry before the cracks were stopped up. All this was done in the spring ready for threshing in the autumn. Instructions for laying such a floor were laid down in some agricultural volumes as late as the nineteenth century, but it is doubtful whether many farmers ever went to the lengths required.

The threshing machine. The origin of the threshing machine is obscure but it is thought that a Scotsman, Michael Menzies, was responsible for the first mechanical thresher about 1732. He proved that his machine was capable of extracting more grain than the **hand flail** method by passing through it twelve bolls of corn, previously threshed by hand flail, and removing another quarter peck of grain. His invention consisted of a number of flails loosely attached by short chains to a horizontal beam, the whole of which was turned by a gigantic water wheel. The corn to be threshed was placed below the beam in such a position as to be caught by blows from the whirling flails. The invention received a good deal of praise and was at that time loosely described as 'capable of giving 1,320 strokes per minute, as many as thirty-three men threshing briskly'. Unfortunately the flails were inclined to break and Menzies' machine was gradually laid aside. More than twenty years were to pass before a different principle was employed but in the meantime other inventors constructed their versions of the mechanical flail. The greatest problem lay with the breaking of the flails due to the violent manner in which they struck against the floor surface. An unknown inventor attempted to overcome this difficulty by arranging his flails in a row along the topside of a horizontal beam, which was given only a half-revolution through a hand crank handle at one end. The sheaves were brought to this thresher and laid out on platforms at either side.

Transverse section of a Savage threshing machine, *c.*1900

A	unthreshed corn	F	cavings riddle or screen
B	threshing drum	G	elevator
C	concave	H	smutter or awner
D	beaters	J	rotary screen
E	straw shaker	K	winnowing machine

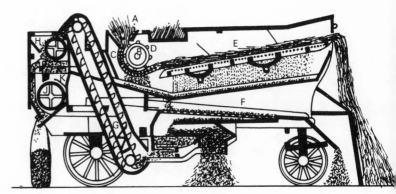

The next two attempts to achieve a practical threshing machine were made separately, but both were similar in principle to the **scutching mill**, used at that time in the preparation of flax.

The first inventor to consider the idea of rubbing the grain free by passing the sheaves between two rollers was a Mr Ilderton of Alnwick, Northumberland. His machine, made about 1776, consisted of a revolving drum some 6 ft. in diameter around which was arranged a number of iron rollers. Each roller was held firmly against the drum by means of a spring device attached to its spindle and the sheaves were passed through. But it seems a large proportion of the grain was bruised.

The principle of this machine was used by another Scotsman, Sir Francis Kinlock, who after some experiment abandoned the idea of rollers working against the drum and instead partially enclosed the drum inside a corrugated iron cover. He arranged four adjustable pieces of wood on the outside of the drum so that they pressed against the inside of the corrugated cover when the drum revolved, and thus the grain was rubbed out of the sheaves as they were passed through. A good deal of controversy took place and continued to do so for a long time between farmers and inventors in regard to the various merits of the flail and the new mechanical threshers. Whilst everyone agreed that a mechanical means was preferable, much concern was expressed with regard to the amount of grain lost through bruising. However ideas continued to flourish and a new direction was taken by an agricultural mechanic named William Winlow, who was encouraged by his employers to design and construct a model threshing machine. For his first attempt he used a series of stampers which acted upon the heads of corn like small hammers, but it failed to work properly. He then gave his attention to mechanical flails and, full of expectation, he built a large thresher on similar lines to Michael Menzies', which also proved ineffective. He did however achieve some success in 1785 with a 12 in. diameter rubbing mill, made on the lines of a **rotary quern**, but he made no provision to deal with long straw and so the heads of corn had to be combed free of the stalk before separation could commence. Another rubbing mill was produced in 1797 by William Spencer Dix. It separated the grain by means of a flat stone with a grooved underside which rotated inside a shallow tub-like container.

Andrew Meikle of East Lothian was familiar with the thresher built by Sir Francis Kinlock from Mr Ilderton's idea, as he was employed to improve it, but finding this a hopeless task he set about constructing one upon an entirely new principle. His version was ready by 1786 and worked so successfully that all threshers constructed after that date were based directly upon it or represented his plan in modified form. The grain was separated from the straw by means of a revolving cylinder subsequently called the threshing drum, which was furnished on the outside with four longitudinal beater bars of wood faced with iron. The drum, rotated by a water wheel, was installed within a circular iron covering, there being about one inch distance between the edges of the beater bars and the inside of the covering. The sheaves were entered, ear end first, between two fluted iron rollers in front of the drum and passed through so that the ears of corn were exposed to the upward blows of the beaters. Thus the ears were struck severely for many times and the grain knocked out before the sheaves were clear of the fluted rollers. Later improvements consisted of adding circular rakes to rake off the straw and reciprocating shakers to separate further the grain which fell down through a mesh into screens or riddles. The number of beater bars was increased when steam power became available, but the principle of the threshing drum remained unchanged. Meikle received an English patent in 1788, but his thresher did not come into general use for ten years, in consequence of which he would have died in poverty, except for timely intervention by a group of gentlemen farmers.

Some northern counties, notably Northumberland, Yorkshire and Lancashire, were employing threshing machines built on Meikle's principle before the turn of the century, and introduction into Wales and the midland counties was spasmodic over the following decade.

In his report on the agriculture of Gloucestershire in 1807, Thomas Rudge observed that 'Threshing machines are found in a few places, the expense being an inconquerable objection to their general introduction'. He went on to give a testimony to their utility by a Winterbourn farmer who stated:

My threshing machine was put up by a Scotsman by the name of Geekie, now living in Bridport. It threshes and winnows about five or six sacks an hour on an average and is worked by eight oxen and requires the attendance of three men, one to hand up to the feeder, a second to move the straw, a third to attend the winnowing and a boy to drive the oxen. A chaff-cutter is attached to the machine which requires a woman when it is in work. My corn is threshed much cleaner than any in the parish by hand, as the farmers have admitted and with regard to comparative cheapness, I can, without going into any complicated calculations, say that I can thresh my corn by the machine at less than half the expense of doing it by hand. The expense, including buildings and attaching a cyder-mill and chaff-cutter is about £300.

The threshing machines so far were permanent installations driven by wind, water or horse power and attended with considerable expense. The early years of the nineteenth century brought forth small hand-powered machines, made by carpenters and the like but none the less capable of threshing various crops at moderate expense. They required two men to crank the handles, two men to feed in the sheaves and another to clear the straw etc. Grain was separated by the action of revolving beaters or in an alternative type of hand machine it was rubbed free by the action of spiked rollers. Inventors experimented with portable threshers mounted upon wheels and with travelling machines that obtained the threshing power as they were hauled to and fro across the field. But in 1830 farm labourers came out in rebellion against the new mechanical methods, their concern being that unemployment would arise on wet days when hand-flail threshing in a barn was the customary job. Disturbances spread as far afield as Yorkshire and Devon, and farmers were forced to guard against attack by their labourers who would have nothing but broadcast sowing, flail-threshing and hand-reaping.

Various horse-powered machines were available for purchase by mid-century. Tasker's portable model cost 50 guineas and threshed seven coombs of wheat in one hour. It required only two horses for power and three men to attend it. The illustrations show the machine packed into its cart ready for transport and also taken down and assembled ready for work. It involved the use of two separate items, the threshing mechanism, usually referred to as the barn work, and the **horse-gear**, or driving power, which stood outside in the yard. The two items were connected together by a horizontal drive shaft, and the horses attached to the radial arms of

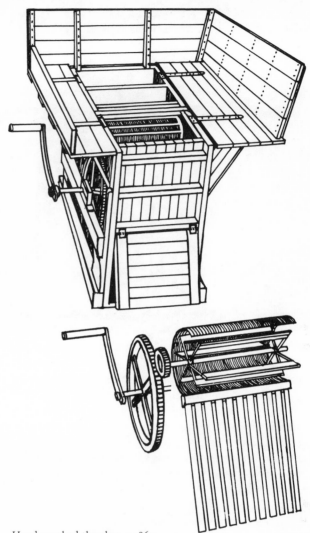

Hand-cranked thresher, *c*.1860
Below: the operative parts: beater bars, concave and slatted screen

the gear stepped round and round continually at a steady rate of two miles per hour. They completed three circuits every minute and their motion, increased to a much faster rate by gearing on the drive shaft, was delivered to four beater bars contained inside the threshing machine. The beaters revolved with rapid motion and corn put in through the top of the thresher was separated with amazing speed, the grains being directed out through a chute at the side, and the straw delivered separately down a grating at the rear.

Portable threshing machines, fitted with winnowers and horse gear attached, were made by various British manufacturers and many were shipped to America and other countries where the

163

TASKER'S Improved Horse Power THRASHING MACHINES.

1867. At the Royal Agricultural Society of England's Show, Bury St. Edmund's a Prize of £8 was awarded to W. Tasker & Sons for their Horse Power Thrashing Machine.

Tasker's portable threshing machine, c.1860

agricultural implement industry was not then fully established. Due to their easy conveyance from one place to another, those farmers who could afford the expense, kept two or three portable threshers of different sizes and hired them out around their district. They usually provided one man to supervise, with the other farmer finding the horse and labour.

Of American inventors, the brothers Hiram and John Pitt of Winthrop, Maine, were the first to produce a combined threshing and winnowing machine. It came into general use after about 1837 and included an endless belt by which the sheaves were delivered to the beating cylinder, where the grain was knocked out. The separation of grain and straw was further completed by a polygonal rake which carried the straw away whilst the grain fell down into a hopper. The chaff was blown away by a fan enclosed inside the machine. The next development in America was the travelling machine which obtained its rotary threshing from the motion of the wheels as it was pulled by horses up and down the field. American threshers differed from the English beater bar type in that they employed a cylinder or drum covered with short iron pegs. The sheaves of corn were first passed between two fluted rollers, which retained them momentarily while the grain was separated by blows from the

pegs, which revolved closely between other lines of pegs projecting inwards from a surrounding iron concave. Most Scottish threshers and small hand-powered corn sample machines employed a peg drum of similar kind. Both beater and peg drum remained, the latter being commonly known as the Scotch machine and chiefly objected to because of the violent manner in which it snatched the sheaves. Charles Burrell and Sons of Thetford, Norfolk, were responsible in 1848 for the efficient portable threshing and dressing machine by which the corn was threshed and cleaned ready for market. Manufacturers were then obliged, in the face of each other's competition, to pay a good deal of attention to the finishing capabilities of their machines and rotary separation screens, riddles and fanners etc. all became standard. By about 1870 the dimensions of threshing machines had been considerably increased in order to accommodate the finishing apparatus and the vast array of gearing and drive straps etc. required to transmit the driving power. They were by far the most complicated pieces of machinery on the farm, often weighing up to 3 tons, and were also costly to maintain. Those farmers who could not afford to purchase employed the service of one of the many threshing contractors who had appeared about the country. The machines were constructed of huge timbers braced from

164

corner to corner with iron straps, mounted upon four wheels and powered by means of a belt from a portable steam engine. Such an arrangement is clearly shown in the delightful engraving of the Ransome, Sims and Head machine, and of particular note is the straw elevator attached to it for the purpose of stacking straw. The use of steam as a power for threshing allowed the drum to achieve a far greater number of revolutions per minute than the former horse power, but the increase rapidly wore down the beater bars and surrounding concaves.

The threshing machine had found its final form as a separate item long before the end of the nineteenth century. Threshing and separation was perfect, grain was cleaned, graded and delivered into sacks at one end of the machine whilst chaff and long straw emerged at the other. The machines were classified as single, double or treble-blast, according to their finishing capabilities. Threshing was very dangerous work, especially for the men who stood aloft and attended the thresher mouth, feeding sheaves to the beaters amidst blinding dust and deafening clamour. In order to relieve this hazard, a Mr Wilders of Grantham invented a self-feeder consisting of four reciprocating shakers to loosen the sheaves after the band had been cut, and a conveyer to carry them into the beaters. Improved self-feeders that came out later included their own sheaf band cutter and a rake to regulate the flow of sheaves fed in by an attendant who stood on the ground safely out of harm's way. Automatic sheaf tying mechanism (a side result of the **reaping machine**) could be attached to catch the straw as it emerged from the thresher and secure it into bundles with twine. **Chaff cutting** machinery was a standard fitment and instead of being trussed or stacked, the straw was chopped into small fodder as fast as it emerged. Improvements were made with regard to the safety and mechanical aspects of the thresher for as long as it was used as a separate item. Manufacturers simplified it a good deal during the early part of this century and adopted the steady drive of an oil tractor. Threshing machines remained essential until about the time of World War II, when combine harvesters that reaped, threshed and cleaned the grain in one operation became widespread.

The winnowing machine. When the threshing machine reached its final form in the nineteenth century it contained a winnowing apparatus which cleaned the grain and delivered it ready for the sack. Before that time threshing and winnowing of grain were distinctly separate operations. The primitive method of winnowing was to pile the corn which had previously been threshed by means of a stick, flail or the feet of cattle, onto a suitable piece of dry land where the grains could be separated from the admixture of chaff by casting both into the wind. The weight of the grains would return them to the ground whilst the chaff and short straw would be blown away. This open air activity could best be practised in hot countries where the weather was reliable, but in England and Scotland a current of air flowing between two open barn doors was found to be sufficient. The idea of winnowing by mechanical means was introduced from Holland during the eighteenth century by James Meikle, the father of Andrew Meikle, inventor of the first successful threshing machine. During his travels in Holland, he had seen such devices, and upon returning to Scotland he proceeded to build a hand-operated winnower which incorporated a series of rotary sieves to contain the corn and four canvas sails to supply the necessary draught. The grain was found to come out free of chaff and so the winnower was quickly accepted in the more advanced districts of Scotland.

An English inventor, James Sharp, tried to improve on Meikle's winnower when he constructed a more compact version in 1770. It worked on the same principle and all the working parts were contained in a wooden casing. It was without canvas sails but employed an internal rotary fan, turned by hand crank handle on the side of the casing. The fan issued a continuous blast of air which was directed at the corn as it passed down through a sieve from a hopper above. The chaff, short straw and dust were blown out at one end of the box whilst the clean grain came out through a screen at the side. This winnower was received with enthusiasm by farmers in Yorkshire and was installed in most of the large farms in that county. Meikle's winnower continued to be popular alongside other models that were founded upon his original idea, and by 1795 many farmers had a winnower beneath their **threshing machine**. Some farmers could not afford such apparatus unless it was purchased collectively. Those who were without made a primitive version of the winnower which was little more than a fan with canvas, wood, or sheet metal sails, to waft air whilst a heap of corn was turned over with a shovel or shaken down an inclined tray. This was known as the 'fan-tackle and chogger'. At the beginning of the nineteenth century, it was customary to place the thresher, which was still a static machine, on the upper floor of a barn, where the grain was separated and then passed down through a winnower to reach ground level free of dust and chaff. The winnower was not much developed beyond this stage as a separate item, but the essential mechanism was included within the framework of the threshing machine when it finally became portable. Consequently the corn was threshed, cleaned and delivered in the one operation.

Hand Tools for Processing Crops

The hand barley hummeller. The earliest form of hummeller was probably a wooden cylinder pressed down and rolled back and forth by hand across a heap of barley. This action separated the awns or beards from the grain and rendered the barley suitable for market or domestic use. A later development, for the same purpose, was the type of hand hummeller that was common during the eighteenth century and which continued to be an asset on smaller farms in the next century. By this time, large mechanical hummellers, either incorporated into the body of the **threshing machine** or separate, were available for use on large farms. Hand hummellers were usually made by blacksmiths, which caused variation to their shape, but, whatever its size or pattern, the principle of its construction and use was identical. The hand hummeller took the form of a helve about 3 ft. in length, furnished at the top with a short cross-piece and joined at the bottom by a socket and iron rods to the hummelling head. This head consisted of an iron frame, between 12 and 18 in. square and 2 in. deep, across the inside of which were set a number of parallel iron bars, each 2 in. in depth and $\frac{1}{8}$ in. thick. The distance between the bars varied with the maker and very often a second set of bars was included crossing at right angles, thus forming compartments, each between 1 and 2 in. square. The objective in each case was to provide a rough, uneven surface with which to strike the grains and break off the awns. This the user did by raising the hummeller and stamping it down continuously upon the heap of barley, occasionally turning over the heap with a shovel in front of a fan, to aid the separation. Hand hummellers were retained by farmers for processing small amounts of barley, whilst many manufacturers were including hummelling machinery within the body of their threshers before the end of the nineteenth century.

The hummelling roller was an adaptation from the above-mentioned instrument, being a number of thin iron discs about 9 in. in diameter, set at intervals of 1 in. along the length of an axle. The discs were joined along their outer edges by thin, horizontal iron bars, also set 1 in. apart around the circumference of the roller to form a cage-like

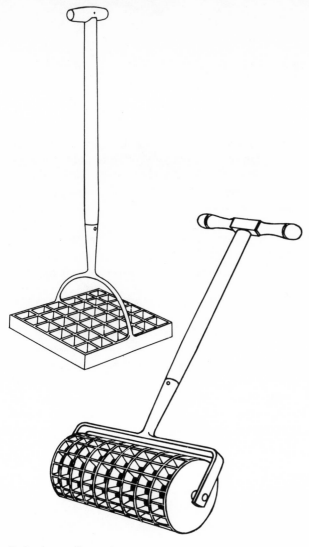

Barley hummeller, c.1850
Hummelling roller

structure. The extremities of the axle were connected by a bow or iron to which was attached a long handle enabling the roller to be held down against the barley and moved back and forth under constant pressure. This instrument was generally replaced by the **mechanical barley hummeller.**

The hand turnip picker. Where turnips were plentiful, sheep were folded in such manner as allowed their access to a limited portion of the crop each day. By such an arrangement they were able to consume the crop gradually, leaving no waste in any area except for the partly buried shells of the turnips, which were later removed by hand with a turnip picker and put through a slicing machine. In addition the sheep consolidated the soil and left behind a liberal quantity of dung.

In 1716 John Mortimer considered sheep's dung to be the best of all dungs, especially for cold clay,

'which not being so conveniently gathered as other dung is commonly conveyed upon the land by folding the sheep themselves upon it'. In this way both dung and urine were saved and as quickly as possible ploughed under.

Folding was not only practised as a means of manuring land for succeeding crops. Frequently the fold was passed very quickly over that which had been newly sown with corn, so that the treading of sheep buried the seed and consolidated the soil around them. This ancient Egyptian practice was superseded early in the nineteenth century by improved harrows.

Sheep were also put into enclosures during the yeaning season when lambs were born. If it were possible the shepherd would erect the hurdles on the sheltered side of a straw stack, forming first an outer boundary wall and a desired amount of subdivisions within, in order to retain the mothers and the young or perhaps to separate the ewe lambs from the ram lambs.

On small farms where the flock was not large enough to warrant a shepherd, a standing fold was usually erected in a dry, sheltered position near to the farmyard. The sheep were enclosed there and large quantities of manure were easily obtained if the fold was littered occasionally with straw, leaves or cavings. The value of the manure depended upon the food which the sheep had been fed; this mainly consisted of turnips which were carted from the field and reduced on a **turnip slicing machine** before being given to the animals.

The turnip chopper was then used to break down the shells or whole bulbs whilst they lay strewn about the ground. Many variations of this instrument were devised, the most common having two sharp blades, crossing at right angles, which were used to cut the turnips into quarters. The two blades were each 8 in. wide and 4 in. deep, crossing each other at the exact centre and were positioned either horizontally or vertically at the end of a stout wooden handle. When the blades were in a vertical position, the labourer who stood with his feet apart, simply raised and lowered the chopper in a stamping action until the turnips were reduced to quarters or smaller pieces by the blades. Horizontal blades were preferred, since they allowed the labourer to use the chopper in the same way as he would use an axe. This action was not as strenuous as with the former type and was less dangerous, as the cutting blades were not then near to his feet.

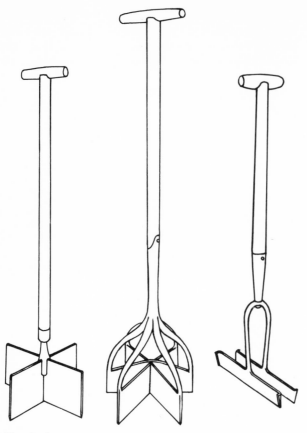

Turnip choppers

3 ft., to be held vertical in both hands and thrust down into a turnip and cut it into slices equal in thickness to the distance between the parallel blades.

The turnip knife. Knives specifically made for chopping off the tops and tails of turnips as they were removed from the soil became available alongside other factory-made instruments during the nineteenth century. However, many farmers made use of their old scythe or sickle blade. With a good half of the blade broken off, the remaining portion of the sickle, nearest the handle, was especially prized for the 'tone' that it had acquired over many years of use. A section, about 12 in. in length, from the middle of a scythe blade was highly recommended when fixed into a short handle. Knives especially made for trimming turnips had a short, curved lever attached to the end of the blade which was of assistance when lifting the bulb from the soil The turnip chopper, slicer and knife have now largely disappeared, except in sheep farming areas. The feeding system does not now depend so much on turnips, so the instruments are seldom used.

Also extra weight was provided above the cutting blades by the addition of two prongs that were used to pull the turnip from a heap or, when necessary, to lever the turnips out of the ground. This attachment was seldom used because, unlike the turnip picker, it drew the tap-root out along with the turnip and these were considered harmful to sheep. Occasionally this instrument was made with three or four blades.

The hand turnip slicer was a simple but useful instrument frequently used when thin slices were required rather than an assortment of pieces. The slicing was achieved by two parallel blades, 12 in. long, 3 in. deep and 1 in. apart. Each was attached by an iron rod from its centre to one end of a wooden handle. This slicer was made to a height of

Turnip trimming knife

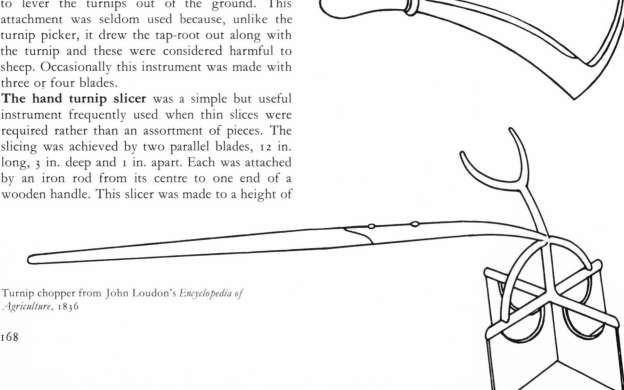

Turnip chopper from John Loudon's *Encyclopedia of Agriculture*, 1836

Grain Drying

Whenever corn was harvested in a damp state it required to be threshed and dried immediately, otherwise the sheaves would turn mouldy and render both grain and straw unfit for anything but litter. A variety of elaborate corn drying kilns became available during the nineteenth century, but until that time farmers who were unfortunate enough to experience a delayed harvest either sustained a great loss or effected the preservation of their grain by use of a heated stone floor. The drying process was carried out in a specially erected building, equipped with a raised floor which was heated from below by a slow fire of peat, turf or charcoal. The grain was spread out over the entire surface of the floor to a depth of 3 or 4 in., and remained there for about twenty-four hours, during which it was frequently turned over in order to prevent scorching. The temperature of the floor was maintained high enough to dry out the grain and destroy any insects such as weevils, which may have been present in number. Some corn drying houses were erected upon the plan of Russian kilns, in which sheaves of corn were stacked for a day or so upon a raised floor of slatted timbers, below which was the glowing remains of a turf fire.

Stack-coolers and grain dryers were promoted by various inventors during the 1870s as being indispensable to the making of good hay etc. Perhaps the simplest arrangement was a hand-powered suction or exhaust fan, connected by iron pipes to the shaft or void at the centre of the stack. This was of particular advantage for controlling the fermentation of grass or corn that had been stacked in a damp stack, and was operated in order to draw cold air into the stack whenever the temperature reached a dangerous level.

There was a similar but more elaborate ventilation system employing underground pipes with branches to any number of different stacks. The large exhaust fan was installed in a barn and driven by a steam engine.

Another plan was to dry the hay or corn in heated chambers before it was stacked. Hot air was drawn from the smoke-box of a portable engine and circulated through the chambers by a fan whilst the damp hay was constantly agitated by revolving rakes. In an alternative system a blast of hot air was forced through sheaves stacked around a wire-mesh frame. Damp grain was dried during its passage through steam heated cylinders, and hay by perforated rollers that emitted hot, dry air. This type of equipment had become too complex for the average farmer by the 1890s and was largely cast aside until its modern revival.

169

The Corn Mill

The corn mill. The early Egyptians threshed their grain and then stored it in deep straw-lined pits, or alternatively, where vermin was common, in containers raised up on stone pillars. When flour was required, a portion of grain was removed from the store and ground down on a primitive hand mill or saddle quern.

The saddle quern, delineated in Egyptian hieroglyphs, consisted of two stones, the lower one being flat except for a slight concave in its upper surface, whilst the second or upper stone was smaller and shaped to fit into it. The Egyptian woman knelt on the floor with the lower stone sloping away from her, and she fed the grains onto the nearest end whilst rubbing the upper stone backwards and forwards over the concave. The grinding action of the two stones reduced the grain to a coarse meal which gradually worked its way down the slope, until it fell off the furthest end onto a rush-matting.

The rotary quern used by the Greeks and Romans made use of two large conical stones, one fitting inside the other; the upper stone was rotated by means of a handle fixed into its side, whilst the lower one remained stationary. The grains were poured through a hole in the top and entered between the two grinding surfaces, after which they emerged from the bottom as fine meal suitable for making into bread. Hand rotary querns believed to date from the third and fourth century A.D. have been found in Sussex and instead of conical stones two flat, circular stones placed one above the other were employed. They were held in place by a central pin and the lower or bed stone remained stationary whilst the upper stone was rotated upon it by means of a handle. Large scale rotary mills of a similar nature were worked by slave and animal power. Their force was applied at the end of a horizontal beam attached to the upper stone, and by walking in a circle round the mill they rotated the upper stone upon the lower and reduced all the grain that passed between them.

During the Middle Ages the English landlords erected water mills to turn their grinding stones and the labouring serfs carried their own corn to their lord's mill, where it was ground into flour. They understandably preferred this to the chore of using hand querns, but it was through this custom that the inhuman practice of the manorial mills became established, wherein the Lord of the Manor enforced his tenants to use his mill and no other, not even their own. The tenant also had to pay a toll in corn for the privilege. The consequences of evading the custom were severe: he was heavily fined and his quern stones taken away.

The rotary hand querns gradually disappeared as the labouring population turned to the miller and purchased enough flour for their requirements instead of grinding their own. However small grinding mills were still required, as in many cases only a small quantity of flour was needed at one time. Portable mills worked by hand were manufactured especially for shipment to the newly developing colonies and for the army, but a large amount of stones were damaged or broken by inexperienced users and the rigours of constant travel.

By the time of Napoleon I, the French had overcome the problem by using metallic discs instead of grinding stones, and the kind of mill employed by their armies on the road to Moscow, was described many years after by R. L. Ardrey: 'It consisted of two circular cast-iron plates about 12 in. in diameter, placed in a vertical position. One was fixed, the other rotated by a hand crank. The plates were indented all over by radiating grooves, and the corn was conducted to the centre, or eye, by a lateral hopper. The meal, as it was ground was projected from the periphery by the centrifugal force of the revolving plate.' At the same time manufacture of flour for the civilian population was increasingly expended by windpower, with an expert miller to oversee the operation. The farmer however was, out of old habit, feeding the animals on whole grain, but meeting with increased prices he was forced to seek improved and less expensive methods of fattening his stock. In order to avoid waste, he purchased a patent **steaming apparatus** and cooked a variety of roots and fodder along with grain, which he found went much further if it were kibbled or crushed into meal. Out of necessity he installed a corn mill of some description in his barn, and if it were a large one, the heavy stones were rotated by horses, but commonly they were small enough to be worked by hand.

By the middle of the nineteenth century, **steam engines** were affording a more constant and regular power for the business of corn grinding and whilst British manufacturers were adapting their mills to

Buckeye combined feed mill and horse power. From
R. Ardrey's *American Agricultural Implements*

it, many American farmers were using a horse-powered feed grinder called the Little Giant. This mill was built in cast-iron with a hopper situated above, and the hopper, along with a stone below, was rotated by horses, who were harnessed to a beam and walked in a circle around it. Corn thrown into the hopper was seized by revolving spiked teeth and pressed down between the stone and its corresponding shell, by which it was ground into coarse meal before falling into a box placed below. A development from this was the famous American Buckeye Mill, produced by H. C. Staver of Chicago, about 1880. It ground feed corn in similar fashion to the Little Giant but also included a horizontal drive shaft, which rotated from the mechanism of the grinding mill and transmitted power to threshing or barn machinery, some distance away. By its use, the farmer could complete two jobs simultaneously and economise on time and labour. By the end of the nineteenth century the common

pattern of corn mill, used in most countries, had two stones each with a grooved surface in contact with the other. One remained dead whilst the other was rotated at a high speed by belt and pulley from a steam engine. The numerous models available at that time were similar in appearance, the main difference being that some utilised stones whilst others were furnished with metalic discs. These machines were about 5 ft. high and the framework was in cast-iron with four stout legs, the whole being surmounted by a large iron hopper to contain the grain. A regulating device was normally located in the base of the hopper and allowed the grain to fall in a steady trickle through the feed spout where chaff and straw were sifted from it by a blower. When the grains had descended they were caught up by a worm thread on the spindle as it passed through the centre of the dead stone and by it were drawn between the two grinding surfaces. The grooves were made somewhat deeper near the centre of the stones in order to accommodate the incoming grain, which then worked itself towards the outside where the grooves were shallower. The distance between the stones could be varied by means of a set screw on the end of the spindle, and so adjustment could be made to suit the kind of corn going through the mill. The farmer seldom required his corn to be ground into fine flour, so he kibbled or cracked it, which resulted in a coarsely textured meal for feeding to animals in either a dry or moist state. Composite grinding stones were introduced in the early twentieth century and were found to be more durable than natural stones. They also wore down quite evenly and could easily be redressed whenever necessary. This type of machinery is rarely found on farms these days, and stones have also gone out of common use in flour mills. Two types of mill used on farms at present are the hammer mill and the roller-crushing mill; the former employs a series of steel hammers rotating at high speed inside an iron cylinder, and the latter, cylindrical iron rollers. In both cases the grain is pulverised, and in some cases the meal then falls directly into an automatic mixing machine where it is blended with other ingredients into cattle food. The resulting mixture may be passed through a pelleting machine, from which it emerges as hard, clean pellets. Grinding corn for bread is now mostly carried out on large mills at specialist factories. The mills work very much on the lines of those described.

The mill-bille was an essential instrument for the periodic dressing of mill-stones. This work was generally performed by skilled itinerants rather than by the farmer or miller, as the task often took several days to complete. The mill-bille as illustrated was about 16 in. in length. The sharp-edged bille, made of high carbon steel, was tapered at each end and wedged into the handle or thrift. The man used it in the manner of a hammer but with a slight sideways action, tapping gently with the object of retaining a level milling surface and a consistent number of grooves to every inch. The grooves were cut gradually deeper towards the periphery or 'skirt' of the stone.

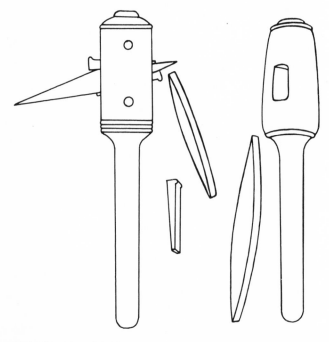

Mill-bille, bit and thrift

Flax and Hemp

Flax was generally prepared for the linen industry by mechanical means following 1835, but until that time and for some years after in the more remote districts of England and Ireland, the fibres were dressed by hand in four separate stages. Firstly, rippling to remove any seeds left after threshing; secondly, breaking or bruising the stems; thirdly, scutching the fibres clean of coarse tow; and fourthly, heckling or combing ready for market. The following instruments were commonly employed for the process and the last mentioned is still used to prepare certain types of reed for roof thatching:

The bindle for threshing or pounding flax, consisted of a solid block of hardwood about 12 in. square and 3 in. thick, the underside of which was cut with triangular grooves and the upperside furnished with a pole handle about 4 ft. in length. The user employed the power of his shoulders to strike the instrument down as hard as possible onto bundles of flax laid out upon a wattle hurdle.

The rippling comb cleared the stems of seed before scutching commenced, and simply consisted of a wooden board with numerous iron spikes set along one edge. It was secured to a post or any convenient upright, and the flax was repeatedly drawn through the spikes in small bundles until the grains had been detached from the stems.

The flax brake or bruising device consisted of three or four triangular planks, each about 18 in. in length and 2½ in. in width, secured side by side across a horizontal framework. A similar number of triangular planks were dovetailed into them, as illustrated. In order to bruise or crush flax, the user laid a handful of dried stems across the lower planks and with his other hand forced the upper planks down upon them repeatedly, whilst gradually moving the stems until their whole length had been treated and the fibres made ready for scutching.

The scutching board stood upright with a niche or half circle about 2 in. in depth cut into its topmost edge. The scutcher whipped a handful of bruised flax against that edge for a brief moment and then repeatedly twisted it back and forth through the niche, at the same time beating it with a wooden mallet in order to remove the bark and other debris from the fibres, before they were heckled.

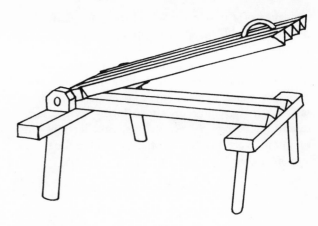

The flax brake

The heckling board was secured horizontally and contained several rows of vertical iron wires about 4 in. in length through which the flax was drawn, first by one end and then the other. The heckler would first lay the fibres parallel and then separate them into parcels of different lengths.

The scutching or swingling mill was the earliest mechanical means of separating coarse tow from flax. It was used in Scotland after about 1780 and within a few years had spread to England. It was similar in working principle to the corn threshing machine of that time, and in like manner consisted of several beater or scutcher bars projecting from all sides of a shaft, the whole of which was rotated very quickly by a crank handle or by a foot treadle mechanism like that of a pole lathe. The flax was held by hand, and each end thrown alternatively against the revolving scutchers until the unwanted bark was broken into pieces and the fibres largely separated. This was the kind of mill used by itinerant Scotsmen who travelled the north of England and swingled flax for different farmers along the way. Mills of identical principle but much greater dimensions were installed in linen factories where a large quantity of fibres were separated for spinning into continuous threads. They were afterwards made much finer by passage around a number of rollers covered with tiny spikes, under whose tension they were pulled out and stretched without being broken. The finished thread was finally wound onto quills and dispatched to the weaver.

A balling mill was employed to separate the fibres from hemp stalks. This was achieved through the crushing action of heavy wooden stampers, with rounded ends, which were raised up and allowed to fall by their own weight into the hollows or concaves of a baseboard where the stalks had been placed. The lifting and dropping mechanism was operated by a man, whilst a boy was continually occupied turning over the stalks in the interval between each stroke. The balling mill, so called after the noise it made, was constructed by mill-wright or carpenter on those farms where large quantities of hemp were grown. Smaller amounts of hemp were processed on a scutching board in similar manner to flax.

7

Barn machinery

Straw Cutting Machines

Straw cutting machines in their most advanced form developed from a simple cutting box.

By the middle of the nineteenth century there were numerous heavy iron machines, designed to cut effectively with knifed wheels and feeding rollers. These were expensive and beyond the means of many farmers, who continued to use the cutting box which was easily transported and involved few moving parts.

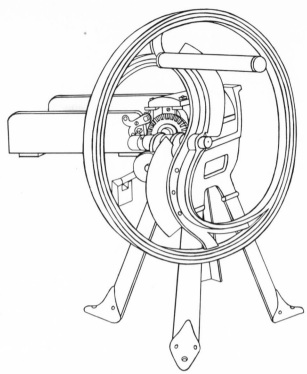

Late nineteenth-century chaff cutter

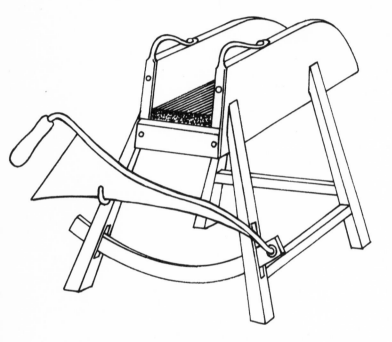

Straw-cutting box, *c.*1760

The cutting box consisted of a wooden trough or box about 3 ft. 6 in. long, open at both ends and along its top. This was supported on four stout legs at a height convenient for the user. Hay, straw and roots were laid lengthways inside the box, and the portion that protruded beyond one end was cut off by means of a knife-edged lever. A device that enabled the operator to hold the straw or other material fast whilst it was being cut was later attached to the box. This took the form of a wooden block positioned inside the box, which was pulled down onto the straw by means of a chain and foot treadle. When the treadle was released, the block raised itself on springs, leaving the straw free to be moved along for the next cut. The operator was aided by a small wooden fork which he held in one

hand to move the straw along, whilst he manipulated the cutting lever with the other. A device such as this would have demanded concentration and a degree of experience from the user.

Before the end of the eighteenth century straw was cut into small pieces by passing between two rollers, one of which was fitted with a spiral knife. A later development in straw cutting machinery was the attachment of sharp blades to the spokes of a wheel, revolving vertically on the end of a trough. Straw pushed along the trough was caused to fan out by passing between a series of vertical iron rods before it was cut into pieces between the descending knives and the bottom edge of the trough.

Both principles were furthered by a long line of manufacturers on both sides of the Atlantic, and a super-abundance of straw cutting machines appeared during the first half of the nineteenth century. They were by this time commonly known as 'chaff cutters', made in cast-iron to a variety of sizes and specifications and mainly designed with an eye to export. One such machine useful to an average farmer is illustrated, whilst most firms offered a larger steam driven chaff cutter to be used where horses, sheep and cattle were kept in number. Such machines were of considerable attraction to the breweries, quarries and railway companies etc. who used horses at that time. Ransome and Sim's for example offered a large steam-driven machine that could be adjusted with precision to cut three convenient sizes of chaff, $\frac{3}{16}$ in., $\frac{1}{2}$ in. $\frac{3}{4}$ in. It was estimated to cut almost two tons of $\frac{1}{2}$ in. chaff per hour.

Worth's American chaff engine came to the notice of the Royal Agricultural Society of England in 1842 and although its principle was opposed to that which had directed the makers of chaff cutters until then, the Society was delighted with its simplicity and ease of operation. They added that this American machine could not be excelled or even equalled in regard to its clean and precise manner of cutting. Richmond's of Salford were responsible for the English distribution of this cutter, which was small in dimension and was intended for the needs of the small consumer, although larger models may have been produced some time later. Its continuous cutting action was achieved by a hand crank handle which worked two rollers held together under pressure with the straw passing between them, and if the operator had a ready supply of material at hand, the motion could be perpetrated until the cutting was complete. The lower of the two rollers had a number of straight sharp-edged blades attached at equal distances apart in the direction of its length, and worked against the upper roller which was formed by a mixture of lead and zinc. This provided a firm bed for the knives whilst not dulling their edges. The two rollers were positioned near to the end of a horizontal trough, along which the uncut straw was moved by the operator's hand, to be caught up and chopped between the rollers.

Chaff cutters did not develop beyond this stage except for some mechanical refinement. Intermittent feed action, direct attachment to the threshing machine, improved forward and reverse gearing, along with a device for sifting the chaff and returning long pieces back to the trough, were introduced. Trials and competitions for straw cutters were arranged almost up to the turn of the century, but the machine had long since found its final form. They remained very important items and were employed on farms whilst horses were kept in number.

The hand sieve for sifting small seeds from corn and chaff

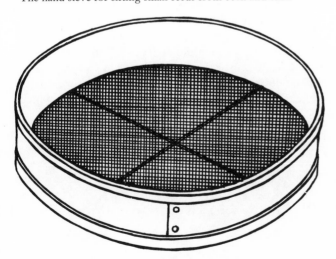

Food Preparation Machines

Steaming apparatus of various capacity became available during the first half of the nineteenth century. Farmers were then able to prepare animal food in bulk, by steaming a mixture of hay, chopped oats and chaff, then adding to it a concoction of boiled roots, kibbled grain and oil cake. Proportions were various, each farmstead having its own particular recipe, often determined by the availability of the ingredients. Steaming ensured that the fibrous parts were rendered more digestible and mixing ensured a more balanced ration for the animal stock, much superior to feeding raw foodstuffs, some of which could be avoided by the animals. Principal items of the steaming apparatus were a steam generator, a wooden tub with iron or copper lining in which hay etc. was treated, one or more iron pans for boiling vegetables, and various pipes, temperature gauges and unions for joining the parts together. The iron vegetable pans were usually cylindrical, capable of holding about six bushels, and hung at either side so as to facilitate discharge of the contents into pots, pails or barrows, by which it was conveyed to the stock. Food steaming apparatus was usually situated in the pulping room, and there a stationary steam engine was installed for general purposes; the same boiler supplied the copper and cooking pans. Farmers, and more particularly their wives, were gradually relieved of this chore in the latter half of the century as a selection of compound cattle foods, in cake and nutted form, became available.

Hand crushing mills for beans, peas, barley, oats and other grains were known but not widely used before the nineteenth century. Their crushing action was controlled by the closeness of grooved rollers and the treated grains used for animal feed. This ancient practice was intended to assist mastication and digestion, so small hand mills, and later, large machines with steam-driven rollers could be found wherever large numbers of working horses were retained.

The oilcake crusher. Oilcake is made from linseed and rape seed after the oils have been extracted. It was first recognised as a manure, then in the 1760s was appreciated as a form of concentrated cattle food, primarily because of its fattening qualities in sheep and bullocks, also because the manure left behind by the animals who ate it was greatly

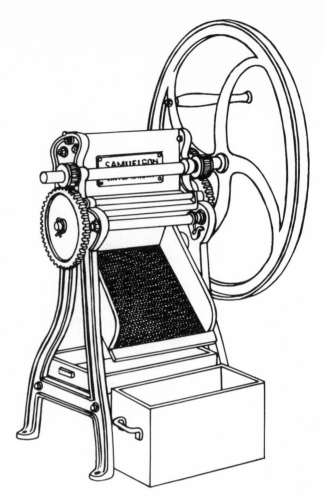

The large flywheel on Samuelson's cake-breaking machine enabled it to be worked with very little power

enriched. Very soon manufacturers set up to provide for farmers on a large scale, and for the convenience of transport and storage they produced oilcake in thin slabs, 15 in. by 12 in. and sometimes larger. But the cakes were so dry and hard that farmers were obliged to install strong machines in order to crumble them down into small pieces suitable for animal feed or alternatively into the powder used as manure at seeding time.

Oilcake crushers were produced in a great variety of styles throughout the nineteenth century, all of them making use of at least one pair of rollers which crushed and split the cake as it passed down between them. The earliest models were fitted with one pair of spiked rollers turned by hand. When steam power was available a second pair of rollers were placed below and each roller then given a greater number of spikes. The Royal Agricultural Society did not offer a prize for a linseed cake crusher until 1845, when they awarded first prize to F. Deans and Co., a Birmingham firm who specialised in bone crushing machinery, used for the preparation of bone manure. By this time, cast-iron rollers about 18 in. long and 9 in. in diameter were generally employed, the surfaces being notched, grooved, fluted or spiked, according to the maker's pattern. When at work, the whole cakes were fed by hand one at a time, between the upper pair of rollers, and the broken pieces could be directed out of the machine at that point simply by throwing an iron shield across the lower pair of rollers. If smaller pieces were required, the cake pieces were allowed to pass through both pairs of rollers and fall onto an inclined gauge or grating by which the largest particles were directed into a basket placed below. That size of cake was reserved for feeding to cattle, whilst the dust and powder that passed through the wires of the gauge was given to sheep or distributed over land through a **liquid manure drill.** The cake breaker continued to be used during the first quarter of the twentieth century, and the convenience of such a machine enabled farmers to extend their selection of cake concentrates which were more economical and easier to store than the nutted variety. Those farmers who grew linseed or flax for the linen market had **scutching machinery** and a press for making oilcake installed on their farms, particularly in Northern Ireland where flax once grew abundantly.

The bone breaking machine. The prime object of installing such a machine was to bruise animal bones into small pieces and less frequently to crush them down into a fine powder. During the eighteenth century bone manure was used for hot-bed and garden culture, but it was not regarded as important and neither was it used to any extent in agriculture before about 1830, since when it has been applied with the utmost success to meadow and arable land, sown either by broadcast distributor or drilled along with the grain. It would seem that the first large bone grinding mill in England was constructed at Hull towards the end of the eighteenth century, at a time when it was a common sight to see women and children scavaging for bones which only found their way to that city for processing. The better bones were put aside for buttons and knife-hafts etc., whilst the remainder were ground down into fine horticultural manure. Small pieces of bruised bone were found to be a better proposition for agricultural uses, and as a

ready supply of fresh animal bones was available on many farms, the owners took advantage of the small crushing mills brought out by numerous manufacturers and installed on their home premises. The 1858 farmers' magazines, concerned as always with farm economy, furthered the practice of bone grinding in an article entitled 'The uses of a dead horse'. After suggesting various uses for the skin and organs of the animal it read: 'the bones are next and these weigh about 160 lbs., and are sold at the rate of 4s. 6d. per cwt., either for converting into knife handles or for making phosphorus and superphosphate of lime. They will not do for animal charcoal, but, ground into dust or crushed into $\frac{1}{2}$ in. bone, they make excellent manures, whilst other special manures, for turnips etc., are made from the blood, flesh and bones combined.' Breaking was affected by passing the bones through a series of iron rollers arranged in pairs and moved by animal, steam or water power. The surface of the rollers was grooved or covered with strong teeth which shattered the bone into pieces as it passed between them. The bones were slowly entered into the machine at first to avoid choking it, with the rollers wide apart, then when the bones had passed once through the mill, the rollers were adjusted closer together and the fragments were broken down further. The resulting bone manure was either 1 in., $\frac{3}{4}$ in., $\frac{1}{2}$ in., $\frac{1}{4}$ in. or powder, the latter in most cases being obtained through riddling the larger sizes.

The valuable superphosphate of lime came into agricultural usage about 1840. Many farmers made their own by pouring sulphuric or muriatic acid upon crushed bones. The bone breaker was not installed much after the first quarter of this century, as fertilizer could be purchased more readily prepared.

The barley hummeller, aveling or awning machine, made for the purpose of removing awns from barley was an invaluable item on barley growing farms. Messrs Grant of Aberdeen were perhaps first to market a portable hummeller during the 1830s. Hummelling machines simply consisted of a wire gauze cylinder about 18 in. in diameter and 4 ft. in length, usually inclined at an angle of about 45 degrees. Each end of the cylinder supported an iron spindle, which passed through its length and projected out through the upper end where it terminated with a pulley for belt connection to the thresher drive or a steam engine. The spindle contained two rows of blunt iron beater arms, which were set on opposite sides to each other and rotated with the spindle when power was applied. When at work hummelling barley, an aperture in the upper end of the cylinder was placed beneath the delivery spout of the thresher and due to the steep inclination the grains were forced to proceed to the lower end from where they were eventually discharged. During their passage they were flung about with great force against the wire gauze, by the revolving beaters, and in this manner the awns were completely broken off from the barley and the clean grain was discharged from an aperture in the lower end. The machine, being movable, could be positioned in the appropriate place to receive direct from the thresher or winnower, and could be pulled aside completely when other kinds of corn were to be threshed. Hummelling machinery was more commonly incorporated within the body of the twentieth-century threshing machine.

In those areas in England and Scotland where gorse, otherwise known as furze or whin, was plentiful, it was gathered in large quantities, but before it could be served to cattle the shoots or spikes had to be thoroughly crushed in order to render them digestible. The cutting and preparation was usually performed by a woman, who was provided with leather gloves, leather apron, a forked stick and gorse pincers.

The gorse pincers were made in iron, with sharp edged jaws and handles about 2 ft. 6 in. in length to extend the user's reach. They were gripped by both hands and a scissor motion applied to cut the succulent but prickly shoots, which were retrieved by the forked stick and compressed into faggots. The stick was replaced during the early nineteenth century by gathering pincers, identical in form to those used for cutting, except that the jaws were not made sharp, but grooved, in order to grip the branches whilst they were cut. **The whin bruiser.** The next stage in preparation, after the gorse had been carted home, was to reduce it to a pulp. This was achieved on a grand scale by the crushing machines developed after 1820, but until that time farmers in England resorted to hand chopping with an instrument known as the **whin bruiser**. The same instrument was also used in Scotland for small quantities, but as gorse grew more abundantly there it was commonly pressed in bulk on a large but primitive mill.

A gorse-crushing or bruising mill of simple form but large dimension was generally used in Scotland before the mid-nineteenth century, and as it was a permanent installation, cheap to operate and could also serve a variety of other purposes, such as splitting beans and breaking cattle cakes etc., it was retained for many years in preference to the newly invented iron machines with spiked rollers. The mill was attended by a boy and could crush a full cart-load of gorse in about three hours, or enough to provide twelve horses with feed for several weeks. It was usually erected in the open and consisted of an old mill stone some 4 ft. in diameter, taken from a windmill and stood upright on its edge. The centre of the stone was pierced by a horizontal axle 10 to 12 ft. in length, one end of which was loosely attached to the top of a wooden post by an iron ring, so that when a horse was yoked to the other end and moved forward the stone would revolve and travel a full circle about 8 ft. in diameter. The circular path taken by the mill was paved with flagstones, and the gorse spread out evenly around the course was turned over occasionally by the boy using a fork until it was thoroughly crushed. This installation was not common after the mid-nineteenth century.

The gorse-crushing machine. The first form of machine for this purpose appeared in England about 1820 and employed rows of square iron teeth, each about 4 in. in length and 1 in. square placed in four rows along a spindle. This spindle, with its projecting teeth, was placed in a wooden box below a hopper and was caused to rotate by a hand crank handle, affixed at one end of the spindle and outside of the box. More rows of iron teeth of similar proportion were placed inside the box and allowed the revolving teeth to pass closely through them. By this means the prickles and stems of gorse that were dropped into the machine could not escape being crushed down to a moss-like consistency, ideal for fodder. Crushing machines of this nature seem to have been installed up to about the middle of the nineteenth century and were probably employed to pulp other vegetables, apples, turnips etc. for animal foodstuffs. Root slicers and pulping machines completed this task more satisfactorily during the second half of the nineteenth century.

The root slicer for cutting turnips etc., down to a manageable size for sheep was developed alongside chaff-cutting machinery. The original turnip slicer was probably a knife-edge lever constructed in similar manner to the **straw cutting box**. Another

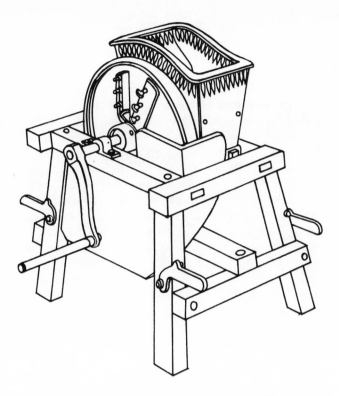

Early nineteenth-century root slicer manufactured by
Mason & Co. of Leicester

number of broad slices, and fell into a basket below. The lever slicer lingered on in most districts but was largely overshadowed by the rotary turnip slicer introduced after about 1820. Some of these utilised two or more knives attached to the spokes of a flywheel, so that bulbs contained in a hopper pressed sideways and were cut into rings or thin slices as the wheel and knives revolved. These were generally referred to as disc cutters, whilst a different form, known as the basket or cylinder cutter, employed a series of blades placed across the length and around the circumference of a roller.

Gardener's turnip slicer was an improvement on the latter class, and the machine as patented by Mr Gardener, an implement manufacturer of Banbury, in 1839, was destined to remain exceedingly popular for almost the duration of that century. The machine stood about 4 ft. high and the framework and casing were in cast-iron. The success achieved by this inventor lay in the manner in which he arranged the cutting knives on the surface of the roller. It was as though a roller with its original pattern of unbroken blades had been sliced into sections 1 in. wide, and each section moved forward slightly so that the cutting knives became staggered towards the centre of the roller in the manner of an echelon. It cut the roots exceedingly well right to the very last slice, thus avoiding any waste. Various

form of level slicer is illustrated. In this model the arm and shaped wooden block contained half-way along its length were brought down with force upon a root and forced it through a series of parallel blades, whereupon it was reduced to a

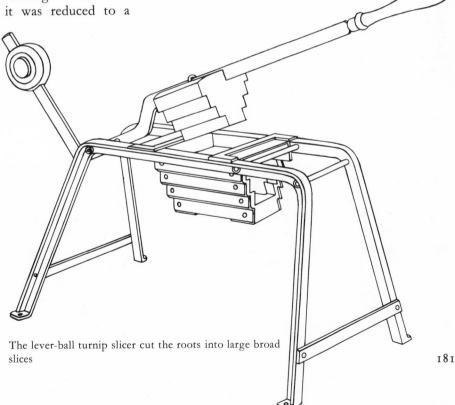

The lever-ball turnip slicer cut the roots into large broad slices

181

other cutting rollers were brought out, and it soon became possible to cut either slices or finger shapes, simply by reversing the direction of the crank handle.

Root pulpers were similar in appearance to the root cutters and also worked on both disc and barrel principle. The knives could be adjusted to shred, slice or produce a coarse or fine pulp as required. The large steam- and horse-powered machines were capable of dealing with 5 tons of vegetable matter, mangolds, carrots, cabbage, etc., every hour, whilst hand-cranked models were suitable for much smaller quantities. When fattening up animals the farmer would use his turnip cutter to a certain stage and then commence pulping food to complete the process. Also young animals and those that were sick soon grew sturdy when given pulped food. These once very commonplace machines are now largely obsolete, but are still to be found on some small farms.

The barrow turnip slicer was perfected by Messrs Ransomes about 1845 and represents the first portable slicing machine to become available for general purchase, although various experiments in the same direction had been carried out by earlier inventors. The farmer no longer had to bring his turnips to the machine and deliver them back to the field or troughs, as the barrow could be easily wheeled from one heap of roots to the next, with the slices either falling into a skep or onto the ground below. The barrow was almost identical to a common wheelbarrow in appearance and construction, except that it was provided with a removable bottom. The slicing mechanism was installed inside the barrow compartment and consisted of a vertical iron flywheel with sharp knives attached radially to its spokes, and a wooden hopper placed at an inclined angle against it. The whole turnips were placed inside the hopper and roller down against the knife wheel, which was rotated by a crank handle. This reduced each turnip to slices $\frac{1}{2}$ in. thick, which then fell through the open bottom of the barrow.

Turnip slicing carts developed out of the barrow slicer and provided a convenient means of distributing slices over a wide area of land in a short time with the minimum of labour. After the leaves and tops of the turnips had been gnawed away by a flock of sheep, the remaining bulbs were lifted with a **turnip picker** and heaped inside the cart, which contained a cutting wheel similar to that on the turnip barrow, but placed at the rear end of the cart and driven by gearing from the landwheels as the cart was moved forward by horses. The slices made their exit below the cutting wheel and were scattered behind the cart in a stream, where they were further consumed by the flock. The cutting wheel was weighted so as to turn with great velocity and could be put into and out of gear. The knives were attached in radial manner to the wheel, and the turnips were forced to roll onto them through the use of a sloping false floor in the cart, which could be removed whenever the cart was required for other agricultural purposes. In some cases roots are still sliced and distributed in a similar manner, the difference being that the modern slicing machinery is mounted at the rear of a tractor-drawn trailer.

The apple mill was extremely important in Devonshire, Somerset and other counties where the manufacture of cider was undertaken. The first stage in converting apples to cider was to reduce the fruit to a pulp, and this seemingly simple operation had to be conducted with the utmost thoroughness in order to achieve a successful brew. A crude machine used in those districts up to and during the first half of the nineteenth century consisted of a wooden roller covered with fine, sharp nails about $\frac{1}{4}$ in. in length, which grated the fruit as it dropped down from a hopper above. It was then reduced to a suitable pomace by beating in a stone trough (the kind used for watering horses) with the blunt end of a pole, following which it was removed and placed in a press, the description of which varied enormously from one district to another.

In most cases, the cider press was nothing more than a strong wooden box about 3 ft. square and 20 in. deep, perforated on all sides with tiny holes. It was mounted on a flat wooden base which projected out beyond the edge of the box, and a channel 1 in. wide and 1 in. deep cut into this projection directed the juice into a receiving jar. Cider was made by filling the box with alternate layers of fresh pomace and clean straw which were compressed by a system of compound levers, or flat, heavy stones, either piled onto the lid of the box or lowered on a centre screw. The object was to leave the residuum completely free of juice, and the length of the operation varied according to the weight applied. The residuum was afterwards given to pigs as food whilst the liquid was allowed to stand undisturbed in open jars for the sediment to collect. It was then put into clean barrels and allowed to ferment, the process of which commenced and terminated in accordance with the room temperature. The duration of this operation was often in excess of one week, after which it was necessary to rack the liquid at least two times, very often more, in order to avoid a resulting brew more akin to vinegar than anything else.

An improved form of apple mill was fitted with two pairs of rollers, which effectively reduced the whole of the apples, including the pips. This was desirable since they added a distinctive quality to the cider. The upper pair of rollers were wooden and were covered by rows of iron spikes which pulled the apples down from the hopper. They were then forced by their own compression to pass between a pair of smooth stone rollers, stationed very close together in order to produce a mucilage suitable for pressing. Cider preparation is now completed in large breweries or factories where sophisticated crushing and pressing apparatus is installed. Primitive apparatus has no doubt survived in some isolated areas of East Anglia and Somerset.

The potato washer. Before a mechanical apparatus was invented, potatoes were washed by hand. They were dipped into a barrel of water, the largest individually and the smallest in baskets. The washing machine appeared during the early nineteenth century along with devices for sorting, harvesting and planting potatoes. Its earliest form was the one that is still often used in the preparation of potatoes for market. A cylindrical wooden cage, equipped with a hinged door, was mounted upon a longitudinal spindle. A handle was attached to one end of the spindle and this caused the cylinder to revolve with its lower half contained in a trough of water. The potatoes were fed in through the door and they tumbled about as the cylinder revolved. The friction of the bars and the rubbing of the potatoes against each other, along with their passage through the water quickly removed any soil from the skins. When they had been subjected to sufficient action within the cylinder, the trough was moved away and the dirty water was emptied from it. The cylinder was then turned until the door faced downwards, thus allowing the contents to fall onto a slotted rack, where they were kept until dry. Later machines for this purpose employed drying arrangements, usually heated fans. This enabled the potatoes to be directed straight into

sacks that were waiting at the end of the machine. They are still very much in use.

The potato separator. It would seem that inventors were slow to give attention to a mechanical means of grading potatoes. One of the first such machines appeared in 1882 and was made by the East Yorkshire Waggon Co. It made use of a cylindrical riddle, approximately 2 ft. in diameter and 6 ft. in length, which was made of iron rods set close together in one half of the cylinder and wider apart in the other. This was attached to a spindle and mounted at a slight angle in a wooden framework. The cylinder could be made to rotate by means of a crank handle on the spindle so that when potatoes were placed inside one end, its movement caused them to tumble along its length. The smallest potatoes fell through the first section where the bars were set close together, whilst the larger sizes moved along to the other section. Here the bars being set wider apart, allowed the middle grade of potatoes to fall through, whilst the largest were cast out through the end of the cylinder. To avoid damaging any size of potato, chutes were provided beneath the different sections of the riddle to guide them safely into baskets. This type of separator is still used, but the various-sized tubers are directed away by conveyors to a bagging device.

The potato riddle, commonly known in Scotland and the north of England as the potato harp, was suitable for small farms and could easily be made by the farmer himself. It did not include any moving parts and simply consisted of two different iron meshes mounted one above the other inside a wooden framework. The potatoes to be separated were placed inside the upper riddle, and those which were smaller than the holes of its mesh would fall through into the second riddle where the mesh was smaller. In this manner the potatoes would be sorted into three sizes, those too large to pass through the upper riddle, those too large to pass through the lower riddle and the smallest size which fell through both riddles into the waiting basket. The upper riddle was horizontal and provided with a high rim to contain the largest potatoes. It was raised some 3 ft. above floor level by the framework to which it was secured, whilst the riddle below was placed at a steep angle like a chute, in order to direct the second size of potato into a separate basket. The sizes of mesh could be varied in accordance with the shape of the tuber and the size of chat and ware required. Factory-made riddles of this kind emerged during the late nineteenth century, and many of them are still in use today.

The smut machine and the similar **flour dressing machine** were developed during the first half of the nineteenth century. The smut machine was a necessary adjunct to the thresher and winnower and, according to an 1850 agricultural report, was the invention of one Hall, late of Ewell in Surrey. This is most likely true because no mention is made of such a machine before that date, probably because it required the speed and velocity of a steam engine to power it. Smut was a common enough disease in corn at that time, appearing as a black soot-like powder in the ears of the plant, large or small amounts being present according to the severity of the disease. It was necessary to remove all traces of the powder before corn was weighed and put into sacks ready for market, so for this purpose the machine made use of a fine gauze cylinder about 4 ft. high and 18 in. in diameter, placed at an angle of about 45 degrees, inside a box. The corn was fed through a hopper into the upper end of the cylinder and whilst descending it encountered row after row of stiff brushes, fixed onto arms radiating from a centre spindle and revolving at considerable velocity against the inside wall of the gauze cylinder. The corn was caught between brush and gauze repeatedly so that by the time it passed out at the bottom of the machine it was clean and shining, all smut, dirt and other foreign bodies having been forced out through the gauze. Later in the century some manufacturers included this mechanism in their threshing machines, and whilst at the same time portable models were available for use in the barn, they continued to remain essential.

Pickling wheat by hand. Before wheat seed could be purchased ready-treated against disease, it was vital for the farmer to pickle his own seed. This task was usually performed in a barn on the evening prior to sowing the seed. The seed was first placed in a cylindrical wicker basket, the whole of which was immersed in a basket containing a weak solution of copper-sulphate. On being removed, the basket was suspended over the mouth of a barrel by means of a rod pushed through the handles and allowed to drain dry, whereupon the grain would be found to be coated with a thin film of copper-sulphate which would preserve the resulting crop from the ravages of smut and bunt fungi.

This method replaced the eighteenth-century custom of soaking seeds in a cask filled with strong brine and sheep dung. Wheat was steeped in this mixture for eighteen hours, barley for thirty-six hours and peas for twelve hours, after which the liquid was drained off and the seed was dried by mixing it with unslaked lime.

Hannaford's wet wheat pickling machine. Smut and bunt fungi were also common in Australia, and it was customary for farmers there to wash their seed in water or to mix it by hand with copper-sulphate powder. This long and tedious process finally came to an end in 1915, when a South Australian farmer, Alf Hannaford, made the first mechanical wet wheat pickler. His machine was little more than a wooden trough about 4 ft. in length and 2 ft. in depth, open at the top and made watertight to contain the pickling solution. Above one end of the trough was a wooden hopper through which the seed was poured into the solution below, the mixture being constantly agitated by a mechanical paddle. When the released spores of smut floated to the surface they were removed by a revolving belt, positioned at the opposite end of the box to the hopper. The belt was scraped clean before it returned to the bath. The inventor followed this eight years later by a pickling machine using chemical dust, which he found to be less harmful to the grain. Farmers now depend upon chemicals in order to protect their grain.

Barn Tools and Utensils

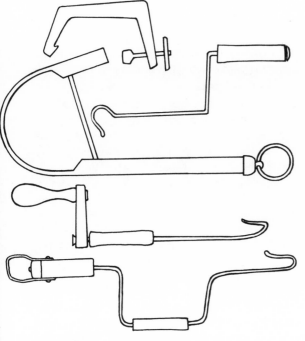

Throw hooks

Gilbert's 'improved' iron sack holder

The throw hook was an early device for making straw rope. It existed many centuries ago when rope of this kind was required in large quantities for tying down thatch, for forming straw drains in the land and a multitude of other purposes on the farm. The throw hook was little more than a bent sapling, cut from the hedgerow, simple but effective even when seen against the **mechanical rope twisters** that were invented in the late eighteenth century. The introduction and versatility of yarn that came later meant a release from the chore of making straw rope, but even so, in some outlying regions, notably in Scotland, this ancient practice persisted into the present century. If it were not possible to obtain a sapling bent by natural growth into the correct shape, then a pliable branch, about 3 ft. in length, bent over and backwards at one end, would suffice. At the opposite end it was necessary to provide a loop of leather or iron, attached to the sapling by a swivel joint. The user would attach it to himself by passing his waist-belt through the loop leaving the hook to spin freely on the swivel joint. To start a rope he would grasp the hook in his right hand and with his left attach some wisps of straw to the terminal point of the hook. Subsequent twisting of the hook on the swivel joint

would cause the stems of the straw to twist together. When a foot or two of rope had been made in this manner it was necessary for an assistant, who sat upon a heap of straw, to retain the end of the rope with one hand whilst laying on measured quantities of straw with the other. As the rope grew in length, so the maker would move backwards, giving his full concentration to twisting the hook and keeping the new rope taut and free of kinks.

A straw rope twister intended as an improvement on the common throw hook became available before the nineteenth century. It was similar in appearance to the carpenter's brace and bit, and it worked on the same principle except that a hook replaced the bit. The procedure when making a straw rope with this twister was almost identical to that of the former implement. The only difference was that the maker used both of his hands to operate the instrument after he had caught up the wisps of straw. The work was just as exacting as in former times since the rope maker had to exercise great care to twist the rope with consistency down its entire length. Over-twisting would cause the rope to snap, and slackness caused the straw to pull apart.

Mechanical rope twisters were produced in abundance during the nineteenth century. The resulting rope was hardly superior to that produced by earlier devices but, with some machines, it was possible to produce up to half-a-dozen straw ropes at one time. It was however still necessary for one person to attend each individual rope as it was being made, but only one person was needed to crank the twisting machine. Normally the machine was stationary, mounted either upon a heavy timber base or set into the ground with the ropemakers working away from it. Its movement was provided by a hand crank handle which caused a series of cogwheels to rotate, these being mounted side by side in a long horizontal box at the top of the machine. Each cogwheel axle was extended out through the front and terminated in a hook, so that when the operator turned the crank handle, each hook was caused to rotate. The rope makers, each with a bundle of straw beneath his arm, would start his rope on the hook before him and as the twisting continued, the maker moved backwards away from the machine, continually laying on more straw which was taken up in the twist. When a sufficient length of rope had been completed, the

maker wound it in a spiral fashion around one arm, thus forming a compact coil, handy for storage and convenient for throwing into a cart or up to the top of a rick when it was being thatched. Throw hooks, hand and mechanical twisters were laid aside as straw rope was replaced by pipes in land drainage, sisal rope became freely available for thatching etc.

The bushel measure was employed in all transactions of wheat, oats, peas, beans and barley etc. It was a measure of capacity, containing four pecks or eight gallons, and could be constructed to any depth of diameter provided that its holding capacity was exactly the same as prescribed by the authorities, who would duly indent their stamp on its surface. Normally it took the form of a shallow barrel, narrower at the top than at the base. It was made in oak and banded by iron, with a fixed handle on either side. It was essential that the mouth of the

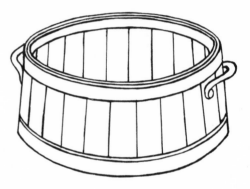

Bushel measure

measure was small enough for whole or part insertion into a corn sack, so that no grain would be lost whilst the measure was being emptied. The Imperial Bushel used in Great Britain contained 2218-19 cubic inches, whilst the American Winchester Bushel contained 2150-4 cubic inches.

Other corn measures used were the peck, which was a quarter of a bushel, and a lippy, which was equal to a quarter of a peck. The corn sack was designed to hold four bushels of grain.

The corn strike was commanded by the Weights and Measures Act as a necessary adjunct to the **bushel measure** and, as laid down by the Act, it was a smooth, wooden roller 2 in. in diameter. It was used to roll across the top edge of the measure and thereby remove any corn that exceeded the level of one bushel. Many farmers and merchants were of the opinion that a strike of this kind rendered an inaccurate measurement by its compressing the corn rather than levelling it. Unofficially they used a flat, sharp-edged strike, drawing it across the bushel in a zig-zag manner, in order to separate the grains and leave an even surface.

The corn scoop, or corn shovel, was constructed entirely in wood and was used to move heaps of grain in the barn or granary. Its short helve, 2 in. in diameter, was formed into a handle at the top. The blade was some 15 in. in length and 12 in. in breadth, with its face made slightly concave in order to retain as much corn upon its surface as possible. It was light enough for a boy to use with both hands, but smaller versions, identical in appearance, were made for use in one hand, leaving the other hand free to open a sack or steady a basket. When the heap of corn had been reduced with the scoop, the remaining grains were gathered together with a small hoe, which like all instruments that came into contact with the corn, was made entirely in wood. Such wooden shovels are still necessary in granaries for loading grain.

The grain sampler was a useful invention brought out by Boby's Company about 1870. With the aid of this instrument it was possible to extract a sample of grain from the middle or the bottom of the heap, which enabled the farmer to judge its condition without disturbing the bulk. The corn merchant also used it to draw off samples of grain from the bottom of a heaped wagon before he committed himself to purchase. It was formed by a thin brass rod 4 to 6 ft. long, with a handle on one end, a few inches of screw thread and a knob at the other. The screw thread passed through a shank which bridged the mouth of a conical container 2 in. in diameter and 6 in. in length, also made of brass. The rod and the container could be screwed apart until halted by the knob on the end of the thread, or they could be screwed towards each other until the knob sealed off the mouth of the container. When a sample of grain was to be collected, the rod was screwed away from the container so that its mouth became open and it was then thrust downwards into a heap of grain to the full extent of the rod. The grains in that area would fall into the container, and were sealed off before withdrawal by screwing in the knob. By this reliable method, the user could be certain that his samples from the bottom had not become mixed with grain at higher levels in the heap. Such instruments are still employed by corn-chandlers.

The wecht, so-called in Scotland and some parts

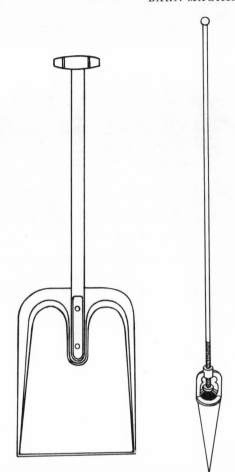

Wooden shovel for moving grain
Grain sampler

of England, was a vessel used for moving corn about the barn. It was normally improvised by the farmer himself, who stretched an animal skin over one end of a wooden cylinder. If so desired this item could have been purchased ready-made, with a light wooden rim and base woven in osier, with a capacity equal to a quarter of a bushel of corn. In those parts of England where the wecht was not common, a wicker basket or a light wooden barrel made from wood veneer was used for the same purpose. The vessel was filled with corn by means of a wooden scoop.

The feed measure. This utensil ensured that horses received the appropriate amount of corn or chaff. It was a cylindrical container with an open top and a handle on either side to assist lifting and carrying. Either wood or light metal was used for its construction and its capacity was usually a half-quartern, a quartern or half-peck.

187

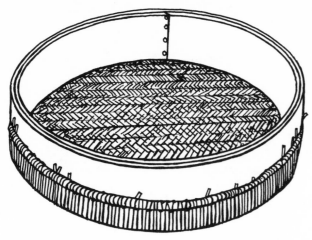

Feed measures were usually made in ½-quartern, quartern and ½-peck sizes

The seed rusky was a basket made to contain seeds and was common to both barn and mill. In the field it was used to supply the sower with seeds from the sack. It was simply constructed, with straw rope coiled one layer above the other and secured in place by osier pegs. The flat bottom of the rusky was raised upon one coil of rope in order to keep the contents clear of the ground. When the rusky was made to a large size, two side handles were provided to aid lifting and carrying, whilst smaller versions did not have handles. The selection of baskets has become limited with the disappearance of rural craftsmen. Large amounts of seed, fertilizer etc. are now transported in paper sacks.

Weighing machines. Most farms were provided with steelyard and scales of some description for weighing straw, grain, wool, and other materials. In addition some had a machine for weighing vegetables by the cart-load and perhaps another for livestock such as lamb, pigs, and bullocks.

The large beam steelyard, as illustrated, would weigh sacks up to 5 cwt. No farmyard is complete without a weighing apparatus of some description. **The hicking stick** was a simple hand barrow used for moving heavy sacks. It was about 24 in. length and carved to shape, as shown. This simple but effective tool no doubt originated in the Middle Ages and remained necessary whilst farmers were commonly obliged to lift heavy sacks of grain. It has been replaced where necessary by expensive lifting tackle.

The hay and straw barrow was constructed in various ways and formed a very convenient carriage for removing sheaves from the stack into the barn for threshing. Its final shape was some variation on a simple wooden platform, furnished at the front end with a single iron wheel and with two handles at the rear for pushing. It resembled the common wheelbarrow in that it stood upon one wheel and two legs, but was without the side and rear retaining walls. For transportation the sheaves were laid across the barrow, butt to ear end alternately, and could be piled as high as was convenient for the labourer. When the rear end of the barrow was raised for pushing along, the sheaves were prevented from falling forward by an upright iron bracket positioned at the front end of the frame. They were essential items in the stack yard before straw-baling equipment was developed.

The hand barrow. One form of hand barrow without wheels was designed for carrying large stones over a distance or for lifting them from the field and tipping them into a cart. It consisted entirely of timber and was made to lie flat along the surface of the ground so that the stones could be rolled directly onto it. Its two main beams were each 6 ft. in length and some 4 in. thick. They were positioned parallel, about 2 ft. apart, and six cross spars connected both beams together. The ends of the main beams were curved up and shaped to form handles by which two men, one at either end of the

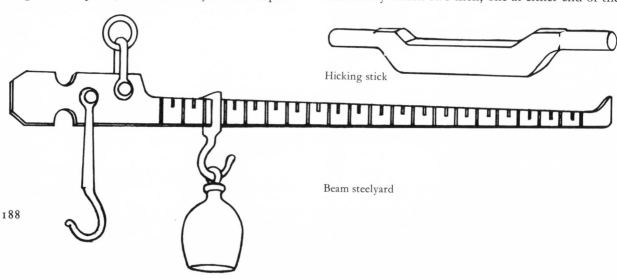

Hicking stick

Beam steelyard

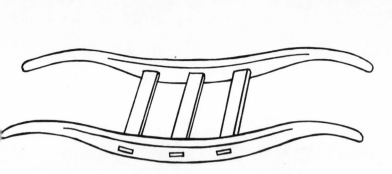

Hand barrow for moving sheaves of corn

barrow, could lift it to a convenient height for carriage. The handles could rest on their shoulders if necessary.

The whin bruiser consisted of a large baulk of timber, usually oak, which stood some 3 ft. 6 in. high and was made to the shape of a large skittle or ninepin. This mass of timber was designed to provide weight above the sharp chopping blades which were attached to its base. These blades were 2 in. deep and were positioned in a parallel manner, $\frac{1}{2}$ in. apart. Constant lifting and dropping of the instrument when it was being used to pulp gorse would quickly dull the edges of the blades, whereupon the 'bruiser' would be shod with a new set. Many farmers who had their land infested with gorse purchased a mechanical crusher when they later became available, or alternatively they obtained a special set of gears which adapted their **chaff cutter** to deal effectively with gorse.

The potato scoop was an advantage when potatoes were scarce, as its use created a great many sets from a small amount of seed potatoes, and each was identical in size and shape, thus being convenient for planting with a drill. The commonest forms of potato scoops came into use at the very beginning of the nineteenth century, when scooping out the eye first came to be talked of. The user held a potato in one hand and a scoop in the other, and upon selecting a suitable position pressed the sharp edge of the scoop into the potato. A half-turn of the handle would cause the blade to cut out a half sphere or set, and it was possible to extract a number of these from one potato, providing care was taken to cut in such a manner that each set had one or two eyes. Such scoops were placed aside as seed potatoes became generally more plentiful.

The common shovel has its helve a little more curved than that of a digging spade. Its square metal head, turned up along both sides and top edge, is a convenient scoop for taking up fertilizer etc.

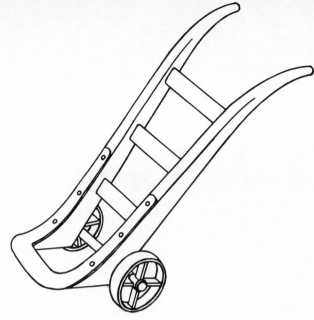

Sack barrow

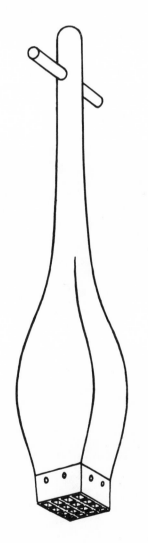

Whin bruiser

189

The besoms used for sweeping hay and corn etc. in the granary or barn were quickly and cheaply made. They were simply bundles of birch cuttings bound with thin strips of willow or hazel to an ashwood handle. Besom-making was once an important woodland industry, but nowadays they are rarely seen.

The malt shovel and **fork** were constructed of plain wood in order not to damage the grain during preparation of malt. They are shown along with a metal scoop which would have been used to remove hops, etc. from the brewing vat. Similar implements are still employed during the treatment of malt.

A grindstone could be found in every farmer's yard until recently. They were particularly used at hay and corn harvest time for sharpening scythes, and the knives of reaping and mowing machines etc. The most convenient size of grindstone for general farm purposes was about 24 in. in diameter. It was mounted in a frame, worked by a crank handle or foot treadle, whilst the lowest part of the rotating stone passed through a trough of water which supplied the necessary lubricant. Small treadle power models were also manufactured for use in the field. They were common until quite recently, but have disappeared along with the many harvesting tools and machines which required constant maintenance and keen edge.

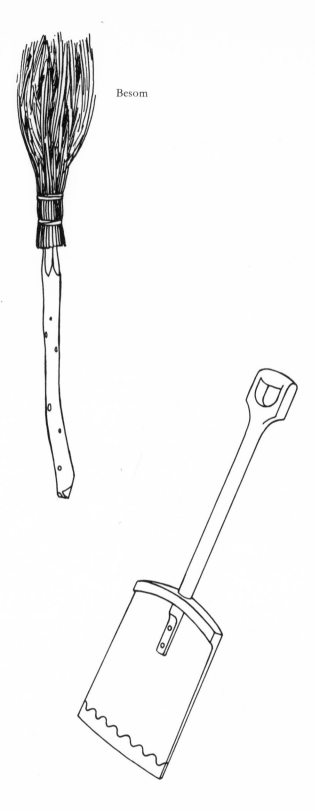

Besom

Malt tools were preferably made of wood, as iron bruised the grain. The metal scoop (*centre*) would have been used to remove hops etc. from the brewing vat

Wooden barn shovel with the handle secured to the blade by two bolts

8

Tools for use with livestock

Before sheep shearing began, the flock had to be washed in order to remove grit, insects and other impurities from the fleece. In some districts this was simply achieved by forcing the sheep to swim back and forth across a deep stream, but the more frequent practice was to construct artificial ponds or 'wash dikes' into which the sheep were placed and washed by three or more men. The first man was equipped with a poy, a long handled pole with a half-hoop of wood or iron at the furthest end, which he used to grasp the sheep and thrust them under water several times before pulling them one at a time towards the second man. He took the sheep by the legs, turned it over, sluiced it, plunged it about, and pummelled the fleece after which it was handed to the last man, often the shepherd, who carefully examined it before it was allowed to swim ashore. The washing operation was usually carried out during the morning in order to give the fleece adequate time to dry and allow the sheep to recover from their ordeal before the chill of evening descended. If the fleece were exceptionally dirty it was washed once again a few days later, but a second washing was avoided if possible as it was considered to damage the softness of the wool. Shearing did not commence until a full week had elapsed, in order to ensure the fleece was thoroughly

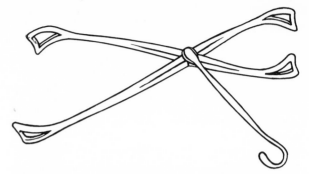

Late nineteenth-century shepherd's obstetrical forceps. Prior to this date, the shepherd would have used his hands

dry. First thing in the morning, the sheep to be shorn were penned near to the shearer, who if he were an experienced hand would complete over fifty during his working day. The majority of sheep presented no difficulty, but the wild and restless mountain breeds had to be bound by the legs prior to the commencement of the operation. The shearer took the animal between his knees and placed it on its rump with back against him, gripping it with his left hand, whilst with his right he manipulated a sharp pair of spring shears and clipped the wool from the neck and shoulders. He rolled the sheep onto its side, and by kneeling upon one knee he secured it with the pressure of his leg across its neck whilst clipping the wool in a circular manner around its body, until the whole fleece was stripped away in a close and uniform fashion. When it was totally removed it was rolled up very firmly by a second person, the outside being turned inwards from rump to shoulder, and bound together by some strands of wool twisted into a rope. The shearers were always neat and expert workers, who endeavoured to clip the wool clearly and near to the skin, not only because it weighed heavier but also because the pile was softer near the bottom. A great deal of care had to be taken in order not to cut the skin, so for that reason the shears were blunted at the point. Large-scale hand shearing is now only carried out in the remote areas of the world, being largely replaced by mechanical clipping. In England it has lately enjoyed a revival, along with other obsolete agricultural practices, for competition and exhibition purposes.

The sheep shearing machines. The first mechanical device for cutting wool fleece was invented in 1868 and subsequently developed by a series of Australians. Before that time the amount of fleece that could be taken over a season depended entirely

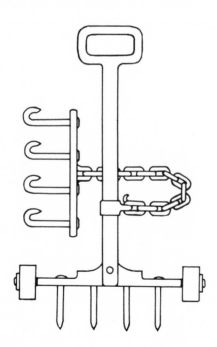

Iron chain used by Rutland farmers for pulling the sheet tightly round bales of wool

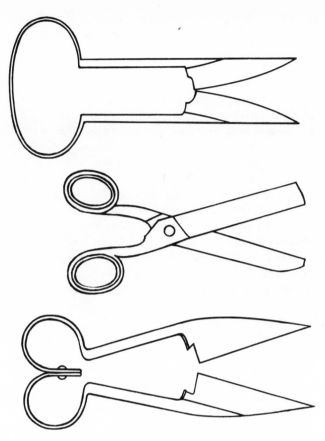

Eighteenth-century scissor shears and nineteenth-century spring shears. Sheep shears were blunted at the points and could be managed in one hand

upon the ability of men using hand shears, and whilst some skilled workers could reputedly blade shear as many as one hundred sheep in a single day, their task was made doubly difficult by having to guide the shears and also maintain a constant clipping action.

James Higham, an Australian inventor, almost provided an alternative in 1868, when he invented a mechanical shear which the operator simply had to guide over the fleece. Had it been practical it would have allowed unskilled labour to perform equally with skilled, but unfortunately it appeared before a suitable driving power was available. However, Higham's machine is acknowledged as the prototype upon which all later shearing devices were based, the essential difference being that he used a miniature turbine as his source of power and located it inside the shear head with a comb and cutter, whilst in later machines a flexible drive shaft was extended from the source of power, usually a petrol engine, and passed to the comb and cutter through the

hand piece of the shear head. Hand shears are still used about the farm for small trimming jobs etc., whilst the mechanical clippers have become more important and increasingly used on all sheep farms since the time of World War I.

The shepherd's crook. The shape and materials comprising the crook have varied enormously throughout the centuries and from one country to another. Ornate specimens of ivory or precious metals have been made but they were mainly used for ceremonial purposes, as the humble shepherd could hardly afford more than the simplest materials for his everyday working crook. However, a great deal of time and thought was given to the design of crooks, no matter what the materials or place of origin, and they were always forged or carved by hand. English shepherds usually had iron crooks which curved round so as almost to complete a full loop, before ending in a knob or tiny spiral. They were leg crooks and used to grasp a wanted sheep by the hind leg which slipped through the narrow opening of the crook and into the loop beyond.

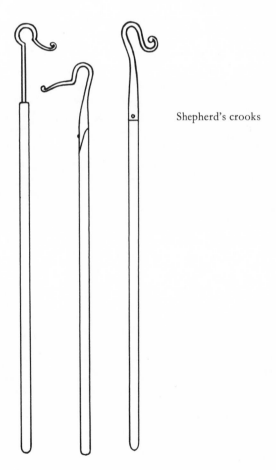

Shepherd's crooks

Scottish crooks were made larger and designed to fit around the sheep's neck. They were sometimes all wood, the shaft and the crook being carved from a single piece of holly, ash or hazel, the latter being most common. The wood was cut when green and seasoned into the desired shape by nailing it to a ceiling beam where it remained for about two years. Otherwise, Scottish crooks had carved horns, fitted on the end of a wooden staff; they too were neck crooks. To fashion the crook, the horn of a ram was first boiled, then twisted and filed into shape, fitted onto its wooden shaft and finally carved with a great deal of sensitivity into a life-like profile, perhaps of an animal or fish. Both types were brought to England by Scottish drovers and can still be found in most northern counties and in Dorset. The crook is seldom seen now except at agricultural shows and demonstrations.

The bath stool was used by shepherds during the eighteenth and nineteenth centuries and served to secure a sheep whilst its wool was bathed for cleansing and medicinal purposes. This stool was usually supported on three legs, each 18 in. high, and its seat was narrow at one end to enable the shepherd to sit astride it. The remainder of the seat was formed by parallel wooden slats upon which the sheep was placed, its head towards the shepherd and its legs thrust down between the slats. The shepherd retained the sheep in this position and at the same time separated the fleece, whilst his assistant poured the bath from a jug especially provided with a long curved spout. It was necessary to turn the sheep over onto its back in order to complete the bathing. The stool was also very convenient when laying on salve, an operation

which took place on all farms where there were sheep. It was disagreeable to the shepherd as it necessitated him smearing the sheep skins with a mixture of tar, butter, fish oil, or other ingredients especially concocted into a paste form. The salve was generally applied during the late autumn for the purpose of protecting the animals against the bitter weather, to remedy scab and kill parasites. The salve was always applied in a particular sequence by the shepherd who opened the wool in stripes so as to expose every inch of the sheep skin for treatment. Each animal took about one hour to salve. This practice was continued by many until the end of the nineteenth century, at which time an order was brought out for compulsory dipping of the entire flock in special tanks or troughs containing powerful disinfectant. This has since remained the standard practice.

The shepherd's salving stool

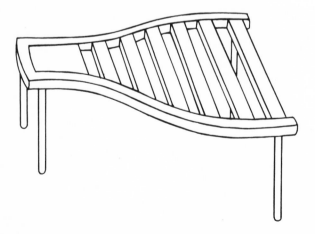

Nineteenth-Century Cow Doctors' Appliances

(1) Nineteenth-century balling guns were used for administering medicine compounded into pellets or balls. Modern drugs can be administered by hyperdermic syringe.

(2) The pressure of accumulated gas in a cow's stomach due to blockage could be relieved with Reade's patent syringe.

(3) **Probangs** for removing foreign bodies such as pieces of potato or turnip, etc. The original probang was a pointed stick. The rigid nineteenth-century stilet probang, (3), was apt to inflict damage to the animal and was generally replaced before the end of the century by (4), a flexible rod covered by a leather sheath.

(4) **Horn trainers** are still occasionally required.

Mouth cramps or gags held the horse's mouth open whilst its teeth were inspected or medicine administered. The gags illustrated were forged in the nineteenth century and replaced wooden ones.

Bells. Sheep and cattle often had a bell hung around their neck on a leather strap, and its tinkling informed the herdsman of their whereabouts. Such bells were mostly made by blacksmiths, being little more than canisters of thin iron with a clapper of iron, bone or any other suitable material. They were made to different sizes, the largest suitable for cows, the smallest for sheep. Bells were circular and mostly cast in brass for horse harnesses. Such bells are still considered essential in less-developed countries by small peasant-farmers, to whom each animal is a very valuable item.

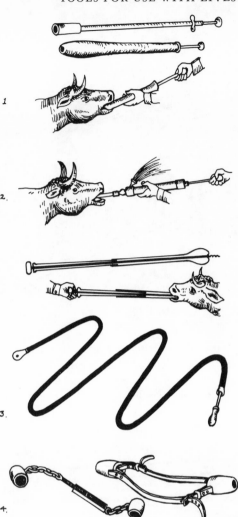

Nineteenth-century cow doctor's appliances

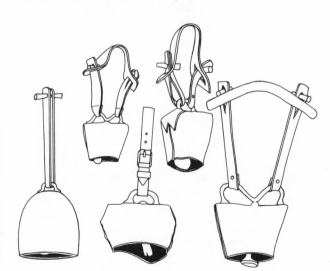

Sheep bells and cow bells

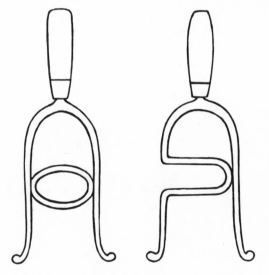

Mouth gags for horses

195

Branding or marking irons

The branding irons for marking animals were forged in iron to withstand heating in fire. It was often the owner's initials at one end of a shank, with a wooden handle at the other for applying the burn. They have now been replaced to a large degree by indelible markings.

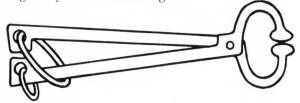

Pincer pattern of bull leader, about 18 in. long

Bull leaders have been made in many different forms for the purpose of leading a bull by its nose-ring. The nineteenth-century patterns, as illustrated, fall into two categories, being either a clasp, hook or spiral at the end of a 3 ft. handle, or smaller hand instruments with a pincer action. Modern bull breeders rely upon a similar pattern of instrument in order to lead their bulls in safety.

The docking iron was used to amputate the last six joints of a young horse's tail. One such is illustrated. It was mounted upon a portable stand and was adjustable so that the horse's tail could rest in the semi-circular notch at the precise point where it was to be amputated. The cutting part was then brought down with a sudden, firm pressure in order to complete the operation.

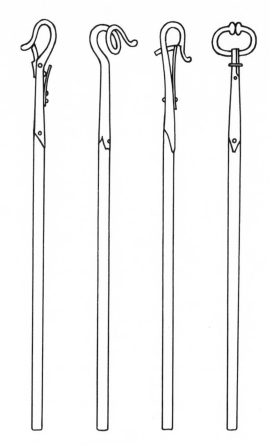

Bull leaders or clip-sticks. When inserted into the nose-ring the bull could be lead quite easily

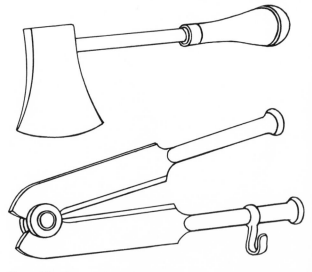

Cauterising iron, 14 in. in length
Castrating shears, 12 in. in length

Hurdle Making

The gate hurdle. Until wrought iron and wire netting hurdles became available in the nineteenth century shepherds enclosed their sheep by willow-wood hurdles, commonly known as gate hurdles. They were made by the shepherd himself or purchased by him from itinerant craftsmen who worked in farming districts throughout the year. The traditional English gate hurdle was made of a suitable size for sheep and was sometimes used to fence cattle, though a more robust wooden hurdle made by carpenters was preferred for that purpose. The task of making gate hurdles in huge quantities called for a liberal supply of well-seasoned poles, efficient tools and no small amount of patience. They commonly consisted of eleven willow poles, two uprights or heads, six rails, a centre upright and two diagonal braces. The manner in which they were put together was first to select two firm poles, 4 ft. 6 in. in length and about 4 in. in diameter, for the end uprights. After the bark had been trimmed they were both sharpened at the butt end and six mortices cut into each to receive the cross rails. These were normally 7 to 9 ft. in length and about 3 in. in diameter, having six inches or so nearest each end made square in order to fit through the uprights, also the three lower rails were placed slightly closer together than those above. Usually the top and bottom rail only were secured to the upright by nails, these being driven straight through the joint. The centre upright, which was shorter than the

This pattern of docking iron was adjustable between 4 ft. and 6 ft. high

Tail docking iron

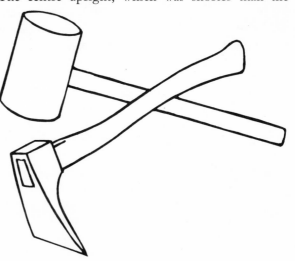

Mallet, adze and driver, the general-purpose tools used in the construction of fences

197

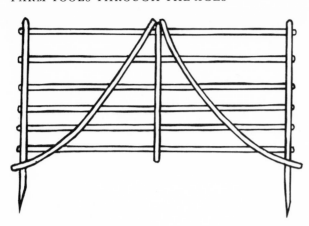

English willow hurdle

outside uprights, was nailed to each cross-piece, and so were the two diagonal braces which extended from either side of the centre upright down to the lower corners. They were slightly bowed to add strength and rigidity to the hurdle. The iron nails used for construction were flat in order to facilitate clenching over at either end. The hammer used by the hurdlemaker had one flat face and the other was fashioned into a chisel-like blade for cutting the mortices. A brace and centre-bit for starting the joints, a gimlet for making the nail holes and a hatchet were the other essential tools of the hurdle-maker. Such hurdles, when being set into position, were laid out on their sides end to end across the field by the builder. He would hold the first hurdle upright on its two points, and by using an iron dibber, make two holes in the ground into which the points of the hurdle were then located. In this manner he would progress along the whole row until all the hurdles were standing upright, and upon returning he would strike the top of each upright with the dibber to ensure a constant height and a firm footing. Before the animals were admitted, each hurdle was linked to its neighbour by means of a wreath, made from slender willow shoots, wound round to form a circle of convenient size and slipped over the adjoining uprights of two hurdles.

The wattle hurdle. The maker's skill was no less required when he made up wattle hurdles. These were of a closer formation than the willow hurdle and were used to form protective walls for exposed sheep and cattle in the depths of winter. To form a wattle hurdle the maker would first place a number of upright willow poles about 18 in. apart, with their pointed ends held firmly in a frame at ground

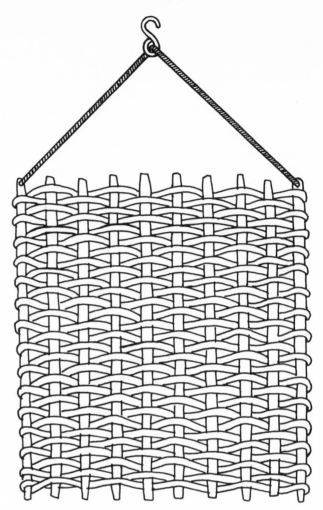

A wattle hurdle was often drawn across soil in order to reduce it to a fine tilth

level. Long willow rods were then split down the centre for their whole length and the halves were woven in and out of the uprights, being twisted over at the end of the hurdle to avoid splitting when turned round and woven back in the opposite direction. When the hurdle was completed from the bottom to the top of the uprights, any rod ends left protruding were trimmed off with a hatchet and the hurdle was lifted free of the frame. If at that stage the hurdle became twisted, the maker would straighten it by the pressure of his knee which was suitably protected by a leather pad.

Sheep netting rather than hurdles was used for folding sheep in some districts of England during the nineteenth century, but generally it failed to achieve popularity. This was mainly due to the problems caused by the weather, as the early twine nets were much given to slackness or shrinkage, therefore needing almost continual attention lest the sheep should escape. In 1842 the Royal

Agricultural Society of England awarded a sum of money to Messrs Wildey and Co. of Blackfriars, manufacturers of sheep netting, in the hope of encouraging the development of a longer lasting article. At the same time Cottam and Hallam of London exhibited a specimen of netting spun from the fibre of the coconut, which appeared to be quite durable when exposed to alternation of weather. Netting was purchased 4 ft. wide in 50 yard lengths and supported by stakes at intervals of 8 ft., with one rope interwoven along the top and a second along the bottom. When establishing a second net the ropes were joined to those of the first, the ends of the two nets were brought together and interwoven with string. Twine nets were gradually replaced by those made of iron wire after the middle of the century, by which time wrought iron hurdles with and without wheels were available for sectional fencing. Most implement manu-facturers included some forms of fencing, both temporary and permanent, in their catalogues at this time, but even so, hand-made hurdles of both gate and wattle type continued into the present century, only to be lost beneath a tangle of barbed wire and the rarity of skilled rural craftsmen.

The fold pitcher was used by a shepherd to pierce suitable holes into the ground for the reception of hurdle stakes. It was not unlike an iron dibber in appearance, being made of the same material, 4 ft. in length and $1\frac{1}{2}$ in. in diameter. The fold pitcher bulged out towards its bottom end to a diameter of 4 in. and then tapered over a distance of 9 in. to a sharp point. This proportion was similar to that of a pointed hurdle stake.

The driver was the alternative instrument used for the purpose of making holes. This was made in oak and was only half as long as the fold pitcher. It was equipped with a point 12 in. long, clad in thin iron plate, whilst the body of wood 4 in. in diameter above the point was strengthened by an iron hoop near to its top. This was to prevent the driver from splitting when it was hammered into the ground. These instruments were extremely useful whenever permanent wire fencing was erected or repaired, and so were retained after the use of sheep hurdles was discontinued. They, and such hand excavating tools as the **post hole borer** and **post hole digger,** were rendered obsolete during this present century by the introduction of earth drilling machinery attached to and driven from the rear of a tractor.

Creels. The Dale farmer often made himself a pair of creels, by which he transported hay to his outlying cattle upon his shoulders. They consisted of a light framework made by bending two hazel rods, each about 6 ft. long into bows or U-shapes, in which position they were secured by rope. The bows were placed about 3 ft. apart and connected together by crossing ropes, so as to form a cage-like structure in which 5 or 6 st. of densely packed hay could be carried. Farmers in the hills of Wales and the northern counties of England used sleds for this purpose, as well as for carrying materials in general. The sled moved on runners and was fitted with horse shafts in preference to ropes. They were especially useful in the winter as they slid easily over the frozen land. Hay-making equipment and motorised transport have largely rendered creels and sleds a thing of the past. But nevertheless they are still much used for transporting hay in some northern regions of Great Britain where motorised transport is difficult.

9

Motive power

Animal Powers

Whippletrees, swingtrees, or swingletrees assisted a team of horses to draw a load of reasonable size with ease, whilst distributing the strain equally.

In early times, the method of attaching animals to a load was one that restricted their freedom of movement and also inflicted a large degree of cruelty. Primitive ploughs were drawn by one pair of oxen with the plough handle tied to the centre of a bar of wood, lashed to the horns of each animal. Some primitives attached the plough by ropes directly to the oxen horns and others fastened the plough to the tail; this barbarous custom is said to have continued in Ireland until as late as 1634.

The Romans yoked their oxen with a collar which was little more than a noose at the end of a rope, and consequently was pulled too tight around the animal's neck. Some improvement was made during medieval times and large teams of waggon horses are portrayed at work in single file, but by then the rigid padded harness had replaced the tight collar which almost strangled the animals when hauling. The new harness took the pressure off the windpipe and enabled heavier loads to be drawn. The earliest representation of a whippletree is on the twelfth-century Bayeux Tapestry. Its use enabled the driver to exert more control over implement or waggon.

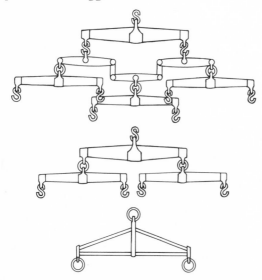

Top: whippletrees arranged for three horses
Centre: whippletrees arranged for two horses
Below: Ransome's draught bars for a single horse were made of light tubular metal

Whippletrees for two-, three- and four-horse teams were a common feature in agriculture up to this present century. They were mostly made in oak or other lasting wood, with iron clasp and eye mountings to receive the trace chains and hooks to join the set together. The common set for two ploughing horses consisted of a main tree and two smaller ones. The main tree was usually about 3 ft. 6 in. in length, 3 in. wide at the centre and tapered to 2 in. at either end. It was attached to the **plough bridle** by an iron clasp and ring at its centre, whilst a clasp and ring were provided near either end for attachment of the furrow and landside trees. They were smaller in dimension but likewise tapered towards their ends, where hooks were provided to receive the trace chains from the horses.

A set of trees for three horses consisted of exactly the same arrangement except that the third horse was positioned in the centre and well to the fore of the other two, at the end of a chain which passed back to the centre of the main tree. The alternative arrangement for three horses was to place them abreast, for which three trees were hooked onto a main tree, one at the centre, the others at either end.

Four horses were attached to a heavy plough by placing one pair before the other, the fore pair yoked by a set of trees, and from the centre of their main tree a chain extended back to the centre of the main tree of the second pair, which in turn was shackled to the plough.

By the middle of the nineteenth century some manufacturers in England were experimenting with whippletrees made with tubular iron, that embodied the necessary strength and offered a lighter weight. Before they were replaced by the single tractor draw-bar there were numerous improved yoking arrangements for the agriculturist to choose from, each having been devised with the objective of dividing the labour as equally as possible between the animals of the team. The arrangement for four horses was still similar to that described, but instead of hooking the main trees to the plough by a chain common to both, the chain was brought back from the fore pair of horses and taken around a pulley, attached to the bridle of the plough, then taken forwards for a short distance to be hooked onto the rear pair. By this method the horses pulled at both ends of the chain, with the pulley caught in its loop and any undue movement of the chain around the pulley would indicate weakness in one pair or the other. Improved yoking for two- and three-horse teams was introduced during the second half of the nineteenth century, and also a compensating device which, placed between the main and smaller whippletrees, would equalise the load and relieve the weaker animals of some strain.

By the turn of the twentieth century every ounce of power was extracted from the horses, but increasing use of steam and then oil tractors meant that horse teams and whippletrees were no longer required.

The ox yoke clearly inflicted a degree of cruelty upon the cattle, as the adjustable neck irons were thin enough to dig deep into flesh. The illustrated specimen is English, of the type widely used up until this present century. It was shaped in oak or other hard wood and only about 4 ft. long in order to keep a pair of oxen close together. The plough etc. was attached by chain and hook to the centre of the yoke. Oxen were eventually replaced by horses, and yokes by **whippletrees**, which ensured a more even load.

Horse-wheel with belt drive to barn machinery

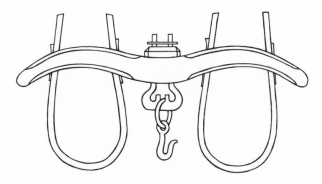

Yoke for oxen

Horse-gear. Before small, compact, steam engines became available, horse-gear of various descriptions provided a source of power for threshing corn, churning butter and other requirements about the farm.

Overhead horse-gear was the earliest plan for applying the power of horses to machinery and was used in England and Scotland during the eighteenth and early nineteenth centuries by those who could afford its installation. Owing to its huge dimensions, overhead gear was necessarily built as a permanent erection within a barn or suitable room put aside for the sole purpose. It consisted of a toothed crown

wheel, anything up to 16 ft. in diameter, which was secured in a horizontal position some 6 ft. above the ground on a vertical post, the ends of which were loosely received into both floor and ceiling of the room. Two horses were harnessed below the crown wheel by a suspended harness, and they were forced to walk round and round in a circle, rotating the wheel along with them. Its regular motion was transmitted to the thresher or other machinery by means of a drive shaft and pinion, the teeth of which geared into those around the circumference of the crown wheel. Horse-gear of this nature was especially the work of skilled wheelwrights, who constructed in accordance with the space available and the nature of the work to be performed. Usually they were placed on the same level as the machinery, otherwise they were on the floor above, but they were always built to remain in working order for a century or more. They fell into disuse as mechanical powers were developed.

Ground horse-gear, usually called the horse-walk or horse-work, was devised early in the nineteenth century and provided a cheap, portable source of power, well within range of the ordinary farmer's pocket. It was made in iron and worked on the same principle as the former overhead gear, but instead of a large crown wheel, it used a toothed drive wheel about 3 ft. in diameter, secured to a strong timber framework lying flat upon the ground. The wheel was rotated by horses, harnessed to the ends of poles, which radiated from it, and the motion was transferred by gearing and a low horizontal drive shaft to the desired machinery, which stood outside of the circle walked by the horses. Whilst the long poles to which the horses were harnessed allowed them to walk a larger, more agreeable circle than before, many accidents were caused by them having to step over the drive shaft between the wheel and the machinery each time they walked the circuit. In order to overcome this difficulty, some farmers placed their gear in a shallow pit with the drive shaft underground. Others covered the shaft with a wooden bridge or ramps, which protected the horses' legs but caused them to ascend or descend a slight hill with each revolution, and young horses were often frightened by the noise of their own tread when passing over it.

By the middle of the century, a number of English manufacturers had developed improved horse-gear, amongst whom were Barrett, Exall, and Andrews of Reading. Their gear was very efficient and safe to use. It was neat in appearance, having almost the whole of its mechanism contained inside a vertical cast-iron drum, 18 in. in diameter and 24 in. high. The horses were harnessed to radial poles, as with earlier gear, and their circular movement was received onto a vertical shaft which protruded out through a covering cap on the top end of the drum.

Each revolution of the vertical shaft was transmitted to a horizontal drive shaft which projected out through the lower side of the drum and went to the threshing machine etc., but in the course of transmission each revolution was increased by gearing, to provide thirty-three revolutions of the drive shaft for each circuit of the horses. The inclusion of a universal coupling joint on the drive shaft allowed the gear to be dismantled, moved about, and reconnected to other machines.

By about 1860, horse-gear had reached the peak of mechanical efficiency and most farmers owned one of some make or another. But by that time, English manufacturers were promoting their portable and small stationary steam engines, which were destined to become the premier source of power during the second half of the century.

The tread power was extensively used on farms in America where it was successfully applied to timber sawing and threshing etc. The engraving (on page 213) of two-horse tread power from the J. I. Case Catalogue requires no further explanation, but for lighter operations such as butter churning, the power of a dog or a sheep was sufficient. Case continued to manufacture their tread mills long after the introduction of **steam power,** but the latter was to prove more efficient in the long run.

Windmills

The earliest windmills were stationary constructions placed on high ground and set to catch the prevailing wind. In the East were the Seistan windmills erected on the Iran-Afghanistan border, with sails mounted on a vertical windshaft to rotate in a horizontal plane. They were no doubt inspired by the horizontal waterwheels known in Asia Minor since the first century B.C., whilst Western windmills were no doubt suggested by vertical waterwheels. It is very likely that some windmills were built on wooden platforms and floated upon water in order to turn about and receive the wind from various quarters. The next development was on dry land and featured the 'post mill' in which the whole wooden structure, containing sails, stones and gearing, was supported and turned upon a massive upright post or column at its centre. This type of windmill had appeared in France and England by the twelfth century, and very soon after they were followed by the 'tower mills', built with either stone or timber, which remained stationary whilst only the cap or dome containing the sails was turned into the wind either by a windvane at the rear or a lever arrangement inside the dome. Such mills could be found all over Europe, where they were employed (in the absence of water to turn a mill all the year round) for grinding corn, or for supplying a cheap source of power for threshing, whilst in the lowlands particularly of Holland they have long been in use for pumping water.

For a short time some Scottish farmers used windmills instead of water mills to drive their threshing machines, but owing to the excessive cost of repair and maintenance, and the dependency upon wind pressure, they were generally discontinued when hand and horse-powered threshing machines became available in the late eighteenth century. Windmill threshers did place themselves very much at the mercy of the weather and could never guarantee their corn being ready in time for market. G. H. Andrews, in his *Modern Husbandry*, told a tale about an unfortunate farmer whose threshing had been delayed and who decided to take advantage of a howling gale, as he badly needed the corn. However when he came to halt, the brake strap parted, and away went the sails without restraint. The farmer became very anxious in case his mill should catch fire or fall to pieces and was therefore obliged to thresh his corn over and over again for a whole day and night until the gale had abated.

The use of windmills, except as a means of pumping water, gradually declined in Europe during the late nineteenth century. However in America, where German and Dutch settlers had earlier introduced the European types, manufacturers of windmills with a different form began to flourish. American windmills differed from the European variety with four sails, in that they were much smaller in dimension and were constructed with a large number of narrow wooden slats or blades in the form of a wheel. The majority of American-designed windmills had the wheel placed vertically, but in some it was horizontal. The wheel was made in two different patterns, first an open wheel composed of a large number of individual slats, all of which could be adjusted simultaneously to catch more or less wind, and second a solid wheel, with the slats rigidly secured to a circular framework, and not folding as did the former. Both types were equipped with regulators to govern their speed and also rear vanes to bring them round into the wind. They were made to various sizes, were perfectly safe during gales and were well adapted to farm purposes, especially pumping water for cattle.

The Halliday standard was a pioneer windmill of the 'open wheel' self-regulating variety. Invented by Daniel Halliday and John Burnham about 1850, it was manufactured by them in large numbers, at South Coventry, Conn., and provided farmers in America with the first reliable means of pumping water for stock and domestic use. In many areas, especially the prairies of the far west, windmills were the only means of raising water from deep wells, and without them stock grazing would have been impossible. The wells were generally a good distance away from the farmsteads but, nevertheless, windmill-pumps maintained a steady, unbroken supply of water without the need for daily attention or the cost of fuel. The important self-regulating device came into action whenever the mill began to run too fast. Water was pumped into a cylinder, and the pressure activated an arm which turned the fan sideways to the rising wind. A reverse took place when the wind slackened, and by this arrangement the mills remained unharmed even in high-force gales. Such windmills are still a familiar sight in many countries, where they continue to pump water most effectively.

The solid wheel windmill was invented by a missionary named Wheeler, who in 1841 had settled in Wisconsin. During his work amongst Indians he encountered great difficulty in grinding enough corn and pumping enough water to provide for the whole settlement, and so considered a windmill as an additional source of power. He was unable to further his idea until 1866, when he constructed a timber model which proved satisfactory in moderate wind but fell to pieces during a heavy storm. With some help from a government blacksmith, Wheeler and his son erected a second windmill, this time provided with a self-regulating device and a side vane which caused the wheel to turn away at an angle whenever the wind became too strong. Subsequent advice from relatives caused Wheeler to patent his successful invention, and large-scale manufacturing was then commenced under the name of the Eclipse Wind Engine Company.

The aeromotor windmill revolutionalised the American windmill industry when L. W. Noyes of Chicago put it onto the market in 1888. It was the result of long research and exhaustive experiment by the inventor, T. O. Perry, who had previously constructed sixty-one different forms of wheel and conducted over 5,000 dynamometrical tests in both natural and artificial wind to determine its most efficient form. As a result he rejected the wooden slats common to windmills a generation before and replaced them with a lesser number of thin steel slats, slightly curved in cross section, which allowed his wheel to respond to the lightest of wind. He mounted it on top of a steel tower and provided the vertical drive shaft connecting the wheel and water pump with gearing so that the wheel would expend three complete revolutions to each stroke of the water pump. This low gearing was necessary to relieve the pump and the open work tower of pressure during high winds, whilst also taking fullest advantage of light and moderate breezes. Every section of the aeromotor windmill, including the tower, was galvanised before assembly and thus received a protective coating to ensure durability.

The popularity of the new windmill was indicated by the rapid growth of its manufacture. Within three years of commencing business on a small scale, the Aeromotor Company had erected their own large building equipped with foundry, galvanising plant and machine tools in Chicago. Their product represented an ideal investment for farmers both at home and abroad, as its use was not only confined to irrigating land and grinding corn, but opened up new possibilities as a means of providing power for driving various light machinery. However, by about 1920, portable oil and paraffin-fired engines had generally replaced the windmill for all tasks about the farm, except that of pumping water.

Water Power

Top: overshot wheel
Centre: undershot wheel
Below: breastshot wheel

The water wheel. Motive power obtained from a fall of water was ideally suited to the purpose of agriculture, owing to its cheapness and reliability. The water wheel originally had buckets or boxes attached to its circumference and was used as a perpetual lever for raising water. They were ancient artisans who first set projecting boards around the rim instead of buckets, so that the vertical wheel could be turned by the flow of a stream and produce enough power to work mill-stones. The importance of water wheels increased over the following centuries and with the inclusion of heavy gearing the power was applied to grinding, pulping and later to threshing etc. There were three categories of water wheel: the overshot, in which the whole flow of water was directed into the wheel by a chute, the breast-shot which was arranged about mid-height in the water and rotated by the force of its current, whilst the undershot wheel obtained its motion from the ebb and flow of tides. For this reason the last was situated in creeks and estuaries where a tide could be retained behind sluices and then gradually discharged to work the mill. In all cases the size of wheel was proportionate to the volume of water, work being started and stopped by the opening and closing of a sluice in the mill-race. Water wheels fell into general disuse when steam engines were introduced, but water power was retained by a minority in the more convenient form of turbines and water engines.

The water turbine depended upon a high column of water in order to produce energy. Briefly it consisted of a wheel with radiating vanes contained inside an iron casing. Water was admitted to the centre or circumference of the wheel, depending upon the model, whereupon it acted against the vanes and rotated the wheel at a velocity relative to the height of the column. Water turbines were never a popular means of powering farm machinery, owing to the high cost of installation.

The water engine was simple and economical but required water piped into it under pressure. In appearance it resembled a small steam engine, with a single cylinder and flywheel from which the drive was transferred by belt to chaff-cutting, or root-pulping machinery etc. Some of these engines were designed to work with both water and steam, the latter being most favoured owing to the general lack of piped water.

Steam Engines

The newly invented steam engines were quickly exploited as a scource of power on the farm, but their use did not become extensive until the second half of the nineteenth century.

Stationary steam engines were fixed to the floor of a barn and employed to drive pulping, grinding and cutting machinery etc. These early horizontal engines, with boiler combined, were mounted across two pedestals, one beneath the fire-box and the other beneath the smoke-box. The boiler was either a cornish or multi-tubular type, the latter being most recommended where fuel and skilled maintenance were limited. At first only large farms could afford to install a fixed motive power, so they were to be found more frequently in malt houses, breweries and pumping stations etc.

Semi-portable steam engines, so-called because the engine and boiler were placed together on the same base, but did not have wheels to move about on. Both engine and boiler were vertical, with the single cylinder secured near the top of the boiler, working a crankshaft and flywheel below. They were often referred to as vertical engines and on the farm had almost superseded the fixed horizontal type by about 1875. They were developed with the saving of time and expense in mind, and became popular because they occupied only a small space in the barn and required the minimum of maintenance. Both stationary and semi-portable steam engines were the main source of power on farms until the petrol engine was developed.

Portable steam engines. Accordingly manufacturers made their engines portable by mounting them upon wheels. A typical example of a portable engine is illustrated. The boiler and engine move together, the boiler forming a foundation for the engine and the whole assembly supported by four wheels. The chimney is shown turned back along the top of the boiler where it remained when the engine was not at work. These early engines, no doubt inspired by the railway engines of the period, were not traction engines and therefore were only of limited value about the farm. They had to be drawn to the work and steered by horses attached to the front wheel-axle. Nevertheless these robust old engines performed admirably and became an essential accompaniment to the portable threshing machines, which required increased r.p.m. for efficient working. It was necessary to provide agricultural engines with some form of spark arrester near to the top of the chimney in order to prevent nearby corn-ricks etc., catching fire. It normally took the form of a circular cage containing conical baffle-plates, and sometimes water, so that it was impossible for glowing cinders to find an exit. In answer to the demand for fuel economy, some compound steam engines were brought out. In these, the steam was further expanded in a second cylinder of much larger dimension than the first and thereby provided a considerable amount of additional power instead of going to waste. Such engines were necessary for driving threshing machines and barn machinery etc. until oil tractors became generally available.

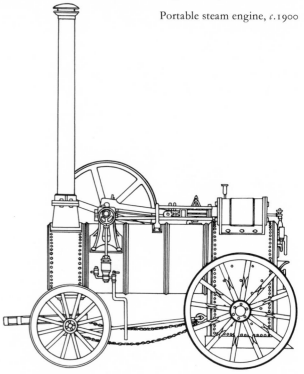

Portable steam engine, c.1900

Straw-burning engines. Some manufacturers fitted their engines with straw-burning apparatus, which was particularly useful in countries where coal and wood were scarce. By means of this invention most forms of vegetable refuse could be used as fuel without any loss of engine power whatsoever. The workings made use of an enlarged fire-box with belt-driven rollers at the door for conveying straw, reeds etc., into the flames. The compression of the rollers caused the material to 'fan-out' as it entered the fire-box, so that each straw caught fire instantly. Messrs Ransomes conducted a series of experiments to ascertain the comparative value of straw and coal as fuel for the purpose of threshing by steam power, and discovered that $3\frac{1}{2}$ to 4 lb. of dry straw would produce the same amount of steam as 1 lb. of best coal in a well constructed boiler.

The discovery of oil in many parts of the world induced some makers to adapt their engine boilers to work on oil as well as solid fuel, but by that time the internal combustion engine, which was destined to change the pattern of agriculture, had appeared.

Agricultural locomotive engines were widely purchased during the second half of the nineteenth century. They were generally found to be well adapted for ploughing, cultivation, and haulage etc. but were not suited to the delicate operation of reaping corn. Therefore farmers were obliged to retain a number of horses in addition to a steam engine. The first public demonstration of ploughing by direct steam traction took place at Louth in Lincolnshire during 1856, when a Burrell-Boydell engine drew four double-breasted ploughs. The engraving (on page 98) shows them at work completing twenty-three acres in the one day. The ploughs used on that occasion seem to have been designed for horse traction, but it was not long before a wide range of implements were available for steam traction. Because of their tremendous power, steam tractors were of particular advantage for hauling the huge ten- to twenty-furrow ploughs required to break up new tracts of land in America and Australia. Steam crawler tractors that did not pack land so much as the wheeled variety were developed after the turn of the century, but about the same time oil tractors made their appearance in farming catalogues.

Gas and hot-air engines were available to the farmer before the end of the century. Each of them was about the size of a small steam engine and intended for permanent installation in a barn. The gas-engine was in some circumstances preferable to steam, owing to greater cleanliness and safety, but the fuel had first to be converted into gas, so its use was not always economical. No water was required for a hot-air engine, the piston being powered by heated atmospheric air. Portable paraffin and petrol engines have been retained on isolated farms, but elsewhere electricity has become the great motor.

Farm Transport

Carts and waggons. Before the advent of oil-engined vehicles, carts and waggons were the farmer's main means of transport and both were built for him by craftsmen according to the local tradition and kind of work to be undertaken. Their history and detail is extensive enough to warrant a complete volume in itself, and I can only hope to introduce the reader to a fascinating subject which began in early times.

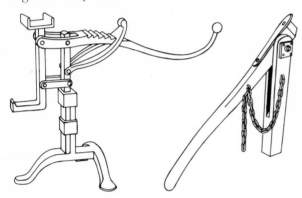

Jacks for raising heavy waggons and threshing machines whilst the wheels were greased

Carts were in use on British farms by about the end of the eleventh century. They were hauled by a pair of oxen yoked by the neck to a centre draught pole, which in turn was fastened to the axle with solid wheels. The body of the cart was simply a platform of rough wooden planking, and the sides, if there were any at all, were fashioned by weaving branches between uprights. Obviously this was the direct ancestor of the tumbril or tip cart which continued as a useful vehicle for most agricultural

purposes well into the twentieth century. Amongst other things it would carry manure to the field, stones to the drainer, farm produce and small animals to market, and hay and corn to the stack yard. In the latter case, sparred wooden frames, or 'ladders' as they were often called, had to be fixed above the sides so that a worth-while load could be carried.

The tumbril was so arranged that by withdrawing a pin or pole, its body was tilted up at the fore-end and the contents discharged behind, whilst the horses remained in the shafts. It was commonly built to carry about 40 cwt. and consisted of a rectangular body with a stout oak frame and plank sides, balanced upon an axle and two wheels, each about 4 ft. 6 in. in diameter.

The majority of tumbrils were fine specimens of craftsmanship and were finely balanced so as not to throw the load too far forward on the ridge tie or horse's back chain, or too far back where it pulled up on the belly band. The tumbril was the most useful cart for general farm purposes especially when its tipping was made gradual, but there were other kinds of cart made for specific purposes also. **The hay or corn cart** was long and low and made for use at harvest time. Generally its wheels were wide and rimmed with convex iron tyres that did not cut and damage grassland as much as the common cart wheel.

The cattle cart was low to the ground for the convenience of loading sheep and pigs etc. Its body was built of light planking and carried by a cranked axle whilst the sides were high enough to prevent the animals from climbing or jumping out. The better cattle carts were finished with movable covers to protect the contents in bad weather, and some had a device to maintain the body on an even level

Harvest cart for carrying hay or corn

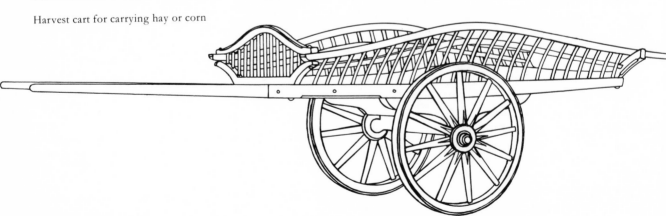

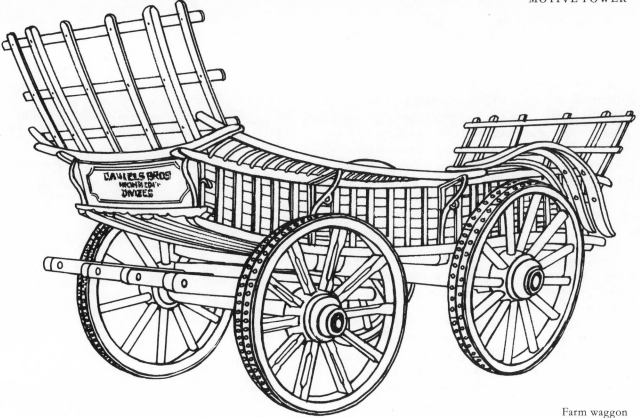

Farm waggon

when ascending or descending a hill. Unfortunately these horse-drawn conveyances have now disappeared from the rural scene, only to be replaced by powerful heavy duty motor vehicles and tractor-drawn trailers.

The pony trap was a light form of cart which the farmer drove to market.

The farm waggon. At one time the cart was universal in Scotland, whilst in England the four-wheeled waggon was preferred to transport farm produce of every kind. It served the farmer well, and besides normal day to day work it removed his belongings from one place to another, and when he died it was painted up and decorated with flowers to convey his coffin to the churchyard. The development of this most useful carriage is shrouded in history. Most probably the sledge was its direct ancestor, the wheels being used in summer and the runners in winter when the ground was otherwise impassable. In later centuries, waggons were hauled about the farm by a pair of horses yoked abreast, whilst heavy loads moving along the public highway often required two pairs of horses yoked in tandem. The rear wheels on a waggon were larger than the fore wheels, the latter pair

being directly attached to the horse shafts and swivelled about in order to facilitate turning the load in its own length. The four-wheeled waggon took a much longer time for the wheelwright to construct than a cart, but at the same time it allowed him to exercise his craftsmanship within the limits of a regional or county pattern. The majority of farm waggons stood about 6 ft. high at the rear, slightly less at the front, and the body generally measured between 11 and 14 ft. in length. The fore wheels were about 4 ft. in diameter and the rear wheels 5 ft. in diameter. Some were tapered or 'waisted' behind the front wheels to allow the wheels to lock round. The Berkshire, Kent and Sussex waggons were in this category, whilst the Suffolk waggon had recesses for the wheels to lock into. The Oxfordshire waggon had curved rails over the rear wheels which emphasized an already elegant shape. Almost every county had its own particular pattern of waggon, and many went to America to become the pioneers of the all-purpose covered waggon.

The wheelwright. A brief mention of the wheelwright must be appropriate at this point. He was the master craftsman of the countryside and began work

in the forest, where the timber was selected, felled
and carried home on a four-wheeled carriage, or the
two-wheeled truck called a 'neb'. There the planks
were sawn out over a pit by hand, until steam
powered saws were available.

Cart and waggon wheels were always made of
wood, and in order to withstand the jarring of
rough roads they had to be made concave or
dished, otherwise the spokes would be knocked out
by repeated bumps. The wheel generally consisted
of a oakwood hub or stock, fashioned by an axe
or turned on a lathe, then bound at each end with
iron hoops. The spokes were cleft from oak or elm,
tenoned into the stock and made to lean out slightly
and receive the felloes or curved sections of the rim.
Finally the wheel was shod with an iron hoop or
tyre which was applied red hot and dowsed with
water so as to shrink tight around the felloes
without warping the wood. The tyre was slightly
conoid in shape and added to the strength and
durability of the dished wheel, in which the lowest
spoke was always vertical whilst its opposite one,
above, leaned out slightly. Most other forms of
horse-drawn vehicle had their wheels constructed
in like manner but, with the coming of pneumatic
tyres, they and the craftsmen who made them have
unfortunately disappeared.

Development of the Tractor

Development of the internal combustion engined tractor began in the U.S.A. and largely continued there. Single cylinder gas engines were mounted upon steam tractor running gear in 1889 and were used for a time by farmers in the North-west. They were not very successful because of ignition and running difficulties during cold weather. As early as 1892 the J. I. Case Threshing Machine Company, then the largest manufacturers of steam engines in the world, produced their first petrol-powered machine, which closely resembled a steam tractor in appearance. It failed commercially because of carburation problems, which were partly rectified by 1895 when Case produced and found a market for several two-cylinder machines.

In Great Britain, Hornsbys and Son of Grantham were producing an 18 h.p. oil-fired Agricultural Locomotive in 1897, and they were followed by Ransome, who offered a 10 ton machine capable of running on paraffin or oil. Like the steam tractors before them, these early motor tractors were beset by faults. They were too large and heavy, with a tendency for the wheels to work themselves into deep ruts whenever the land was soft. Very little progress was made in tractor design before the end of the century, as makers were simply content to replace the steam unit with a petrol or oil engine. They were of the opinion that power and satisfactory performance could not be achieved without a great weight.

In 1902, John Froelich of Iowa produced the first of what was to become the John Deere line of tractors. It was powered by a 20 h.p. single cylinder petrol engine made by Van Duzen of Cincinnati, and incorporated a friction clutch and forward and reverse gears. It was rugged enough to withstand seven weeks' threshing work on a farm in South Dakota, during which time it powered a large J. I. Case threshing machine. The British tractor industry was pioneered about the same time by the three-wheeled petrol-powered Ivel machine. It was exhibited at the 'Royal' in 1903, by which time production in America had accelerated, and various manufacturers were then producing motor tractors alongside their other types of machinery. In 1905 the first factory devoted entirely to the production of tractors was erected in America by Messrs. Hart and Parr. During the early years of the industry purchasers were obliged to convert their horse-drawn implements and vehicles to suit the new power, as those made for use with steam haulage were far too cumbersome. New manufacturers entered the field. Henry Ford built his first 24 h.p. petrol-engined tractor in 1906, having produced a steam tractor for direct ploughing some twenty

Thresher powered by horse treadmill. From the J. I. Case catalogue, *c.*1855

years earlier. The International Harvester Company began experimental work in 1905, which resulted in an ungainly three-wheeled machine with a single wide wheel at the front. In the following year they put their first tractors, built in conjunction with the Ohio Manufacturing Company, onto the market. Their succeeding line of Mogul and Titan tractors became popular and well used in many different countries, but they were still of considerable tonnage and the design very much influenced by the old steam tractors. In fact the Titan could be converted into a road roller by having four 15 in. wide rollers inserted instead of its front wheels. These ungainly machines were the designers' answer to the farmers' requirements, especially those farmers with vast acreages in the west of America, whose measure of work was beyond the capabilities of larger horse teams.

The idea of gaining extra buoyancy and traction by using endless tracks along each side of a vehicle was published in 1825 by an Englishman, Sir George Caley, and again in 1850 by American Gideon Morgan. But it was not until the present century that a crawler or 'caterpillar' tractor was made to employ the principle to advantage. That was Benjamin Holt's steam-powered track-laying tractor of 1904, in which he incorporated separate clutches to transmit the engine power and control the motion of each track independently. This 1891 patent has since remained the underlying principle of steerage on all track-laying vehicles. In 1909 Foster and Hornsby produced a huge steam-driven track-laying tractor which was trialed extensively before military authorities in 1910 before it was shipped out to the purchasers, Northern Lights Power Company in the Yukon. But little interest, outside of the army, was shown towards track-laying tractors in Britain, so Hornsby's gave over the rights of manufacture to Holt's Company of California.

It was World War I and the accompanying shortage of food, man and horse power that did most to hasten the production and general introduction of tractors on both sides of the Atlantic. When confronted by such deficiencies, and no immediate prospect of relief, farmers were obliged to seek alternative powers in order to cultivate an even greater acreage. Tractors were seen as the obvious answer. The unforeseen wartime demand stimulated a good many ideas and developments in design. There were new people in the industry, some of whom were experienced automobile engineers, and they helped to bring forth a trend towards smaller, faster, general purpose machines suitable for the average farm. Before the end of the war a British Government order for 6,000 Fordson tractors was dispatched to America, where, by the use of newly devised assembly line techniques, production had been almost doubled. In spite of the high purchasing price and risk of inflammable fuels about the farm, agriculture was becoming increasingly dependent upon this new power and for the first time it was possible for a farmer to dispense with his horses. Tractors were increasingly used in day to day running and for ploughing and cultivating etc., whilst there was a drive pulley on the side of the engine to power threshing and barn machinery. Most of them were designed to run on cheap, low grade fuels, but in most cases the engine had first to be started and run for a while on petrol, until it was hot enough for the main fuel supply to be switched on.

By 1918 an extensive range of tractors was available. They ranged from small single cylinder units to large four cylinder models capable of hauling some half-dozen binders simultaneously across flat prairie land. All working parts of the machine, engine, gearbox and transmission, were being enveloped within iron casing, chain drive had disappeared and column steering was a standard feature. Measures were taken to determine the most suitable size and weight of iron-rimmed wheels for various soils and climates. Pneumatic tyres were not generally introduced until the 1930s. The most significant development before 1920 was the pioneering by the International Harvester Co. of power take-off, a feature which permitted direct transmission from the tractor's engine to the implement mounted or drawn at the rear. It thus eliminated the necessity for the operative parts of a harvester or mowing machine to work from the power of its own wheels. The new introduction quickly brought on revolutionary changes in the design and construction of farm machinery. They were geared for a higher working speed at greater capacity, made lighter and more compact and became an integral part of the tractor. Power take-off was generally accepted by all manufacturers by the mid-1930s.

Crawler tractors were further developed in America during the 1920s and 1930s in order to meet the demands of forestry and construction work, whilst at the same time their additional power and traction proved to be necessary for hauling heavy combines and multi-furrow ploughs across the vast prairie lands in the West. Combines have since become self-propelled, but crawler tractors have remained necessary wherever the going is too steep, soft or sticky for an ordinary wheeled tractor. They have proved invaluable for bush clearance and deep-drainage work and are commonly used for drawing huge 40 ft. wide disc ploughs on large tracts of land in America, Canada, Australia, and Russia. Smaller countries seldom require equipment of that dimension.

Walking tractors, developed almost concurrently, were characterised by the operator walking along at the rear. He effected control by holding one or two long handles that extended back from where the engine was mounted slightly to the fore of a large pair of driving wheels. A coupling was located near to the wheel axle so that hoeing tines, plough shares, harrows etc., could be attached and drawn along. The operator could observe the work before him, and when correction was necessary he had only to reach out and touch the throttle or clutch control on the handle.

Four-wheel drive tractors are now becoming common and are able to work in conditions that defeat the two-wheel drive tractor. A wide range of specially designed implements are now available for mid- and rear-mounting, and those with wheels can be adjusted to suit the width of the rows of plants along which the machine is working. Fore and rear loaders are now included as an integral part of the machinery.

10

The farm dairy

The use of dairy produce can be referred back to ancient times, and it would appear that sheep and goats were the first animals to be milked.

Excavations of ancient ruins in Egypt have produced evidence of an extremely advanced knowledge of dairy farming, also references to the use of milk and butter are made in *Genesis*. Cattle were kept in number by the Greeks and Romans and they established dairy farms on the outskirts of the cities. As Europe became more civilised, dairy produce provided an important trade link between cities. Individual families provided their own milk, and milk produce, from a single cow and this practice continued up to the start of factory production. Early pioneers of the dairy industry were the Dutch and Swiss, and even to this day they share a leading position in the development of modern dairy methods. In America the dairy industry was started with the arrival of the Jamestown Colonists and their cows in 1607. As cities expanded in Europe and America during the following two centuries, the inhabitants were separated from their farmlands. Early in the nineteenth century large herds of dairy cattle were reared to provide the cities with fresh milk from the country, and any excess milk was made into butter and cheese to meet a growing demand. About this time merchants started to purchase fresh milk, farm-made butter and cheese, eggs, etc. for resale in the cities. This kept the farm dairy alive in the face of competition from newly developing dairy factories. Experience proved that the best butter or cheese came from those dairies where process was entrusted to one individual, so the farmer's wife usually took care of the butter and cheese making. She worked in a spotless dairy using a few traditional utensils, a wooden butter churn of some description and a primitive stone cheese press. They were simple aids, in some cases retained well into the twentieth century, and their development over the centuries is described in the following pages. Butter-making remained a farm process for a longer period of time than cheese-making because it did not require the daily attention of the buttermaker, the cream being saved in a jar each day and churned once a week.

The first cheese factories were established in America by 1831, and they quickly followed in Europe.

Refrigeration was made possible during the 1850s in America and this allowed dairy produce to be moved over longer distances, especially by the growing railways. Other contributions were made to the dairy industry by Gail Borden, who patented condensed milk in 1861, and Carl de Laval of Sweden, who invented the centrifugal cream separator in 1878. Stephen Balcock developed an important test for the butterfat content in cow's milk at Wisconsin in 1878, and Louis Pasteur discovered a method of destroying harmful bacteria in cow's milk. It was made practical in 1895. Dried milk and processed cheese came soon after, and by 1900 the factory production of all dairy products was firmly established, taking almost all of the business away from farms.

Butter-Making Utensils and appliances

A mid-nineteenth-century farm dairy, as described by *The Popular Encyclopedia*, consisted in its most perfect form of a milking-room, a churning-room, a cheese-making room, a loft for drying and storing cheeses and a scullery common to the three first-named apartments, in which the various vessels used in the different processes were washed. In a less perfect form the churning and cheese-making were carried out in the same apartment. The following refers to a dairy of the first class, but is also applicable to one of the second class, so far as the parts described are common to both. The milk-room looked to the north, and its floor was sunk 3 ft. below ground level, but the soil was kept from touching the walls by a face-wall built round it at some little distance. The floor of the space between the face-wall and the wall of the building, as well as that of the milk-room itself, was formed of smooth flagstones, carefully jointed and dressed, and having a channel formed to convey the water away to a drain. The face-wall was 1 to 2 ft. in height and coped with turf and the earth was mounded up to it. In the middle of the milk-room, running parallel to the long walls, there were two benches of polished stone, hollowed out so as to form large but shallow troughs, for holding the **milk dishes** which were surrounded by hot or cold water as required. In the centre of each of these benches was a *jet d'eau* to cool the air in summer; in the winter the water was heated. The benches were supported by pillars, but the space below was of little use and as dirt was apt to accumulate there, it was better to build up four walls to support each bench. The floor of the dairy was not level, but made to slope towards the wall, where a channel was formed to convey the water away and into a drain. The walls of the room and the ceiling were smoothly plastered. In place of a cornice at the junction of the walls and ceiling, there was a simple curve running them into each other, and the skirting, which was of stone or slate, was flush with the plaster, the object in both cases being to avoid any projection where dust or dirt would lodge. In place of a lath and plaster ceiling it was better to substitute an area of brick or stone in the form of a semicircle. The windows of the milk-house were double and the outer pane was covered with

Transverse section of a late nineteenth-century creamer

perforated zinc or wire gauze. At the end of the room, under the windows and close to the floor, was an aperture for admitting air; this was also covered on the outside with zinc or wire gauze and on the inside with shutters. In the ceiling there were apertures for ventilation; they communicated with a well constructed flue, which was carried up close to the boiler flue but was entirely separate from it. At the end of the milk-room was a sink, into which milk dirted by any accident, and the drippings of the milk-dishes, were poured. No food, either vegetable or animal, was ever allowed to enter the milk-house, not even the cream jars were admitted. A popular method of purifying the atmosphere of the milk-house was to dip cloths into a solution of chloride of lime, and then to hang them up on cords stretching from one corner to another. In a similar way the temperature of the room was kept low during hot weather. The other rooms of the dairy were on a higher level than the milk-room, from which the dairyman ascended to them by a flight of steps. They were all, except for the cheese loft, finished in a similar way to the milking-room. In the cheese-making room there was a bench of polished stone for setting out the utensils, built into the wall and supported on polished stone brackets. The churning-room resembled the cheese-making room in every respect, except that provision had to be made, in the case of power-driven churns, for a shaft or belt, which entered through an opening in the wall or an underground tunnel. The floor of the scullery was formed like that of the milk-house, flag-stones sloping to a channel, but the skirting of the wall projected out to afford a better protection for the plaster against large and heavy articles that were moved about. A small aperture next to the scullery contained a steam boiler, from which a divided pipe communicated with both the large washing vessels in the scullery and a cistern placed in a loft above the milk-room. This arrangement allowed the water in the washing vessels to be speedily raised to boiling-point and also provided a constant supply of hot water for washing. The cheese-loft was smoothly plastered and had a closely jointed floor of deal-boards. The cheeses were kept on shelves or better still in the **tumbling rack** invented by **Mr Blurton,** which had only to be turned one half-revolution on its pivots in order to reverse all of the cheeses. The roof of the dairy projected far over the walls and was supported by brackets or by posts so as to form a shed in which the utensils could be exposed to air.

The following extract is from an Agricultural Report on the Vale of Gloucester and Berkeley, dated 1809, and describes the manner of making butter in that area.

The milk, as it comes from the cows, is strained through a hair-sieve into the tin vessels, which are about 4 in. deep. It is allowed to stand only 12 hours, when the cream is taken off with the skimming dish and put into the cream vessels, and the milk is warmed and carried to the cheese tub. The cream is shifted into fresh cream vessels once a day, and is also stirred frequently during the day with a wooden knife, that is always kept in each of the cream vessels. This continued shifting and stirring of the cream prevents a skin from forming on the top of it, which is injurious to the butter.

In summer or in hot weather, several gallons of cold water should be put into the churn, and allowed to remain an hour to cool the churn, before the cream is put in. The cream is strained through a coarse canvas cloth kept exclusively for this purpose, and then put into the churn. The operation of churning should, in summer or in hot weather, be very slow, otherwise the butter will be very soft when taken out, but in winter or in cold weather, and particularly in frosty weather, the churn should be prepared for receiving the cream, by putting hot water into it and allowing it to remain for half an hour to heat the churn. The operation of churning should be performed quickly, and now and then the air, that escapes from the cream in churning, should

be let out of the churn, or it will make the cream froth, and lengthen the process of churning very much. When the butter is taken out of the churn, it is customary with most people to wash it with cold water before salting it. This is never done in Gloucestershire, as the butter retains the sweetness much longer when no water is used in making it up. When it is taken out of the churn, it is well worked with the hand, which presses out most of the milk, it is then beaten out with a cloth, or rather a cloth is repeatedly pressed down upon it, which absorbs all the remaining milk. When this is properly performed and no trace of butter milk remains, it is salted to the taste with finely powdered salt, which is well mixed with it by working it in with the hands. It is then weighed into half-pounds, and made up into rolls about 9 in. long. The process of making butter from the cream of whey is the same as that just described. Butter is made twice a week during summer, and the quantity of milk butter made on this farm is about 16 lbs. per cow, and that of whey butter about 25 lbs. per cow per annum. About $2\frac{1}{2}$ lbs. of salt are used to a cwt. of butter.

The instruments available to the dairyman for testing milk were the thermometer, the cream gauge, the lactometer, the lactoscope, and the pioscope. The first three mentioned were simple to use and so were accepted into most farm dairies whilst the remainder were more appropriate to dairy factories.

The cream gauge was extremely useful for measuring the quantity of cream present in a sample of cow's milk. It was simply a cylinder of glass about 1 in. in diameter and 10 in. high, standing upon a foot. It was usually graduated evenly in $\frac{1}{10}$ of an inch, from 0 to 100 parts, each indicating one per cent of cream. The sample of milk to be gauged was poured into the glass immediately upon being taken from the cow, and after standing for about 12 hours the engraved scale clearly showed the percentage of cream to rise above the milk. Any number of these glass tubes were placed vertically, side by side, in a wooden rack and fixed to the wall of a dairy so that cream from one or a number of cows could be observed during all seasons and in different foods.

The lactometer, probably invented by a Frenchman, M. Quevenne early in the nineteenth century, was similar to an ordinary hydrometer and furnished with a scale to indicate the specific gravity of milk.

The lactoscope, invented during the 1870s by a Professor Feser of Munich, was an optical test that showed the percentage of fat present in a specimen of milk. Several versions were quickly put onto the market, perhaps the simplest one being that developed by a Dr Bond of Gloucester. *British Dairy Farming* of 1885 described it as a 'circular glass dish, on the underside of which was a fixed pattern of black parallel lines clearly visible through the transparent material'. A measured quantity of water was placed in the dish, and the milk to be examined was dropped into it from a special dropper and stirred until the eye could not distinguish the number of lines in the pattern below. The number of drops of milk required for this distortion was then related to a table provided with the apparatus, and the percentage of butter fat easily determined.

The milk refrigerator. As more people moved into towns and cities during the nineteenth century, so it became necessary to transport milk in bulk from outlying farms. In order to prevent this milk from turning sour, it had first to be reduced to a temperature as near to 50 degrees Fahrenheit as possible. This was achieved by a milk-cooling device, loosely termed 'refrigerator', which was mounted upon the wall of a dairy at a suitable height to allow a tin-lined milk churn to be placed beneath it. The temperature of the milk, which was about 90 degrees Fahrenheit when taken from the cow, was effectively lowered by pouring the milk through the refrigerator, where it flowed over the surface of horizontal pipes through which cold water was constantly passing. The illustrated Lawrence capillary refrigerator of about 1850 consisted of a funnel about 2 ft. in diameter into which the warm milk was poured; it then passed through a pipe and regulating valve into the cooling area below, which was formed by ten or more iron pipes, each some 2 in. in diameter and 2 ft. long. These pipes were positioned horizontally, one above the other, and each pipe was connected to the one that lay immediately above it by a small link pipe at one end. Cold water, delivered under pressure into the lowest pipe, was forced to climb through this system until it reached its outlet on the top horizontal pipe. The movement of the cold water ensured that the milk was cooled by the time it had flowed over the pipes and entered a receiving trough at the bottom. It was then directed through an outlet into a churn below. Coolers were eventually made of various sizes capable of processing

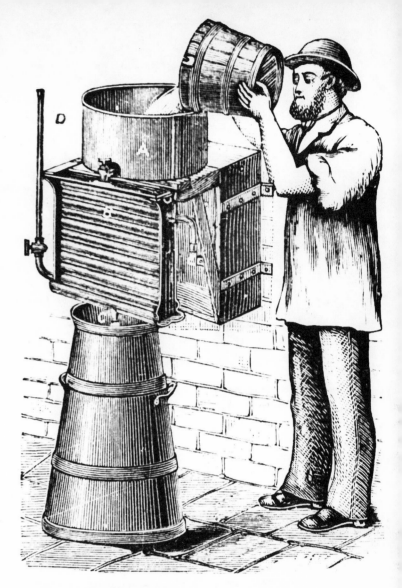

Lawrence's milk refrigerator

up to 500 gallons of milk every hour. They have remained essential dairy items, using chilled water to effect the cooling, and are now built in most dairies as part of the standard equipment.

Milk churns. New railways provided the opportunity for moving milk cheaply and quickly over long distances, so metal churns were employed to contain it in bulk. They were generally made of about 10 gallons' capacity in tin or steel bound with iron hoops, with a substantial bottom and double lids. The original design of a vertical cylinder with lifting handles on each side has hardly changed since. Many dairymen trundled such a churn from door to door on a two-wheeled perambulator or in a horse-drawn cart, measuring out milk with a scoop into a purchaser's jug.

Bottles had long been used in Paris by 1885 to convey milk to the consumer. The first milk bottles

221

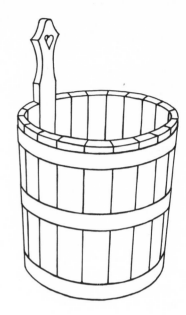

A nineteenth-century milk churn
Dairyman's delivery can with assorted measures
An early nineteenth-century wooden milk pail

were sealed with a cork or rubber bung, which tended to sour the contents. This problem was solved in 1883 by the Lipmann glass stopper, which closed against a rubber ring and the mouth of the bottle by a sprung wire device. It was used until the modern wide-mouthed bottles were introduced, a feature which allows for easier cleaning and cheap, efficient sealing.

Milking pails were generally made of maple, because it was a light wood and clean in appearance. They normally held about 6 gallons each, a capacity which was maintained when galvanised iron and steel pails were introduced about the middle of the nineteenth century. In order to carry two such pails full of milk, at the same time, the milkmaid wore a wooden **yoke** across her shoulders.

An American milking pail called 'the Perfect' was sent over to Europe during 1879. The lid of the pail folded down to form a seat for the milk-maid, milk being directed into a funnel, strained through a filter, then conveyed into the pail by a rubber tube. This pattern of pail was recommended to British farmers, as it was not likely to be kicked over whilst the cow was being milked. Both wooden and steel milking pails were generally cast aside as mechanical milking gained favour. Large and small capacity pails have however remained essential for general purposes about the farm dairy.

The dairymaid's yoke was a beam of wood shaped to fit across the shoulders and partially around the neck. It was generally between 2 and 3 ft. in length and tapered towards each end, where it was fitted with a chain and hook. The user, with the yoke placed across her shoulders, stooped down and hooked each chain onto the handle of a full pail, then by standing upright the pails were lifted clear of the ground, their weight supported by the yoke. It was necessary for the user to grip the chains with her hands in order to steady the buckets and prevent the contents from slopping about. The yoke was common to most countries until recently.

Back cans were especially made for the purpose of bringing milk down from high pastures or trans-porting it across rough country from distant herds. They probably replaced leather and goatskin bags and were made near cylindrical, except that one side was flat, in order for a full can to fit comfort-ably against the farmer's back or a horse's side. They were provided with double, close-fitting lids and handles for emptying. Their carrying capacity was about 7 gallons. Such containers are still occasionally made by local tinsmiths and used in steep districts of England.

The milk sieve. It was necessary to sieve the milk in order to remove any small particles of hay, grass or hair that could have dropped into the pail whilst

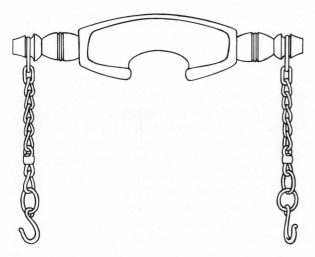

Dairymaid's yoke

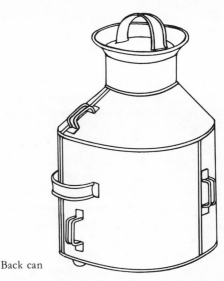

Back can

milking was in progress. This sieve was made as a bowl, about 12 in. in diameter, the bottom of which was formed by a gauze of brass or silver wire. The bowl was normally made of wood, but occasionally stove-enamelled iron and zinc were used. The latter material was at one time used to make a variety of dairy utensils but then discontinued on the assumption that it contaminated the milk and cream. Straining cloths were often used in addition to sieves.

The cream skimmer was normally made in porcelain or tin, of convenient size to fit one hand only. It was used to remove the layer of cream which formed on milk in the setting dish, and was often shaped, with a straight lip to catch up the cream, and a slight concave to retain it. The concave was usually provided with numerous fine holes through which skim milk that happened to be taken up with the cream, could drain away before the cream was placed in a cream jar. Wooden skimmers and scallop shells were common on the Continent and required a good deal of dexterity to use. The cream skimmer was common in every farm dairy and in many country cottages before the present century.

A cream urn was used to contain the skimmed cream until enough had been collected for churning. These urns varied in dimension, but one 24 in. in height and 9 in. in diameter was customary in the average farm dairy. They were made in stone or earthenware, raised up on a base with a moulded handle on either side and a removable top which was pierced through the centre with a hole about 1 in. in diameter. This hole was raised up around its perimeter so that a fragment of silk or muslin

could be stretched tight over it and secured with a loop of twine, thus allowing a small amount of air to enter the urn.

The creamer can was a cheap alternative. It was about the same capacity as an urn, but made in tin, with handles and a lip for pouring. Neither of these items is generally obtainable today.

A cream skimmer

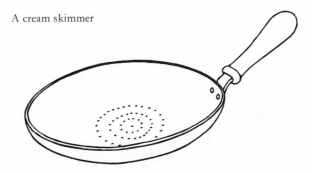

Milk setting dishes. When milk was intended for butter-making it was necessary to separate and remove the cream which is the essential ingredient of butter. For this purpose a shallow vessel or setting dish made in glass, porcelain, earthenware, wood or tin was used. Although its material of construction varied according to the farmer's pocket, all setting dishes were between 2 and 4 in. in depth, and about 18 in. in diameter. This flat form was purposely employed in order to present a large surface of the milk to the effects of the cool atmosphere, and caused the fatty globules of cream to rise above the milk. The cream would have formed after about twenty-four hours and it was then removed by the use of a **cream skimmer**,

223

which left the skimmed milk only in the dish. This was the method of the small farm dairies, whilst the larger dairies and the dairy factories employed large troughs made in marble. Other means of setting the cream, such as steel containers with a jacket of ice or cold running water, became available, but were most appropriate to the dairy factories.

A favoured form of setting dish used in the north of England until recently was a shallow wooden bowl, lined with lead. It had a hole bored through the base which was 'stopped' with a bung whilst the milk was setting. At the appropriate time the bung was pulled so that the skim milk could be drained off into a pail below, leaving the cream adhering to the lead, whence it was removed with a wooden scraper and placed in the cream jar. Setting dishes and other small items such as cream urns and skimmers etc. disappeared as farm dairies were obliged to close down.

Deep setting devices came into favour about 1880 and quickly superseded the shallow pan as a means of raising cream from milk. The old method had been reliable and promoted quick rising, but at the same time had exposed large surface areas of cream to a germ-laden atmosphere. With the deep setting method, milk was contained in deep receptacles below water level, which kept it pure and threw up a large quantity of sweet cream to produce better quality butter. Dairy farmers welcomed this new idea, as it allowed them to handle a greater quantity of milk in a shorter time whilst demanding less space and attention than setting dishes. Another advantage was the larger butter yield which preserved for a longer period.

The Cooley American creamer was a deep setting device available in Britain by 1885. With this model the milk was poured through a strainer into metal cans, 20 in. deep and 18 in. in diameter. Each can was filled to within 1 in. of its brim and then covered with a loose-fitting lid, after which they were packed side by side within a slate lined box. The top was fastened down, the box was filled with cold water, which entered at one end and left at the other, and whilst it completely covered the cans the pocket of air between the underside of the lid and the top of the milk prevented the water from entering. This was an important feature of Cooley's design, as he said it allowed the odours and gases given off from the milk during its rapid fall in temperature to leave the can and be absorbed into flowing water,

whilst at the same time effectively sealing the milk from the harmful effects of the atmosphere. The size of this creamer was varied to suit the size of different dairies. Nine cows required three 4-gallon cans, whilst eighteen cows required six cans, and so on. Cooley was not the originator of the deep setting method but he did improve upon the earlier **Swartz system,** which was named after the inventor and introduced into England from Northern Europe a few years before. Both shallow and deep setting were superseded by centrifugal separation during the early twentieth century.

The de Laval cream separator had been known for some time in Denmark and Sweden before being brought into England to gain its first medal at the Royal Show in 1879. It caused a great deal of excitement owing to the quick and effective manner in which it separated skim milk from cream, compared with the customary method of setting milk in dishes. De Laval's separator secured more cream than the former method, and this was proved by passing through it a quantity of skim milk already separated from the cream in dishes, and obtaining from it a good deal more cream. For the average dairy farmer this machine was somewhat impractical, although there could be no doubt of its efficiency, when installed in a larger dairy or factory, with regard to savings on labour and space. The machine was developed by the inventor and others so that within the decade there were a variety of separators available, the smallest dealing with a few gallons of cream and worked by hand, the largest driven by steam and able to cope with some 300 gallons or more per hour. Very little change has since occurred, and from de Laval's first conception came a number of appliances used in present day industry, and also the modern cream separator.

The illustration shows a sectional view through a cream separator of about 1890, built wholly in steel and mounted upon a wooden base, driven by steam or horse-gear, and standing some 40 in. high. It is without the steel or aluminium discs that a number of makers arranged one above the other inside the chamber to increase the speed of separation.

The whole milk was placed in the heated container above the level of the machine and maintained at a temperature of some 90 degrees Fahrenheit, which was found to aid the separation of cream from the skim milk. It then passed down into the

and upon reaching the top was flung out into the upper compartment, to be released through an outlet. Dairy operatives had no doubt about its great space-saving value, but its chief merit was speed and the dispensing with ice and lengthy cooling processes. The de Laval separator continued to win all the major awards, followed closely by the Neilson-Peterson creamer, a machine from Denmark. In the final decade of the nineteenth century the list of competitors was extensive and included, from England, the power-driven Victoria, Alexandra and Turbine Cream Separators. Manually-operated creamers were manufactured by the Aylesbury Dairy Co., Lister, The Dairy Supply Co., Freeth, and Pocock and Burmeister and Wain. They had become more elaborate and effective by the end of the century, the trend being away from empty bowls towards those fitted with plates or discs which assisted the separation a great deal. The **new American butter separator,** made by the New York firm of Wahlin, separated and produced butter in a continuous operation. It was exhibited in England in 1896. Huge power-driven separators were installed in dairy factories, along with churning, sterilizing and bottling machinery before the turn of the century.

The plunger or dash churns were the first vessels to be designed specifically for producing butter.

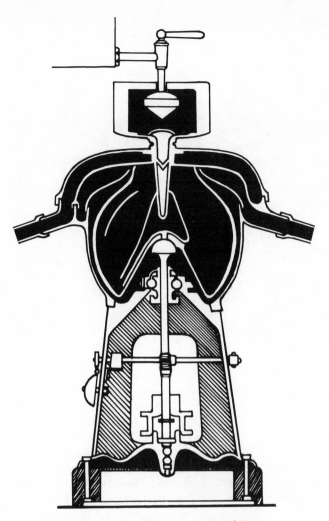

Sectional view of a nineteenth-century steam-driven cream separator. An endless belt turned the driving pulley

Plunger churns

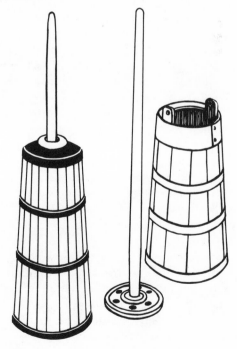

receiving cylinder at the top of the machine, which regulated the flow of milk by use of a conical float. From there it was released through a strainer and passed down into the globular container. This container revolved at a considerable speed, some 4,000 r.p.m., being driven by a vertical shaft and horizontal pulley to a steam engine. During separation the heavier skim milk collected around the sides of the chamber with the lighter cream having gone into the middle. The rising skim milk was caught between double walls inside the top half of the chamber, and from there it was passed out into the lower of two compartments formed by the iron casing around the moving parts of the machine. Meanwhile the cream would cling to the middle of the chamber and rise up around the delivery pipe, by which the whole milk first entered the separator,

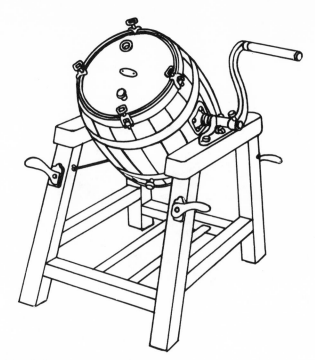

The 'Royal Triangular' butter churn made at Haverfordwest. Queen Victoria ordered them for her farm dairies

Barrel churn

They were in common use by the fifteenth century in Europe and had replaced the time-honoured method of producing butter by agitating milk in a jar. The plunger churn consisted of a stoutly built cooper-work barrel which was always of larger diameter at the bottom than at the top. It was provided with a close-fitting lid, which was held fast by iron clasps or a screw thread around its circumference. A wooden shaft passed through a hole in the centre of the lid at the bottom of which, enclosed in the churn, was a circular piston or plunger, perforated with small holes. The capacity and motivating power of the plunger churn varied, but the common form could be worked by a woman who, after having half-filled the container and replaced the lid, grasped the plunger shaft in both hands and rapidly moved it up and down. This action caused the plunger to agitate the cream and eventually produced butter. Mechanical devices for working this type of churn became available early in the nineteenth century, and they ranged from hand cranked, to horse and steam-powered systems. Large dairies found it profitable to position these churns side by side, and to create a motion by means of a steam-driven, reciprocating lever, which was attached to the plunger shafts and so raised and lowered them simultaneously.

Barrel churns were in use by the beginning of the nineteenth century and, together with box churns, they gradually replaced the earlier plunger variety. The earliest form of barrel churn was made to revolve longitudinally by means of crank handles at each end, and in order to agitate the cream the inside of the barrel had to be furnished with four or more diaphragms, comprised of open slatwork through which the cream could partly flow. For loading and unloading, a bung-hole was provided and whilst it was adequate for that purpose, it was otherwise impossible to inspect the interior of the churn or to cleanse it properly through such a small aperture.

Improved barrel churns, commonly called tumbling churns because of their end-over-end motion, were developed in about 1880, though the former kind continued to a limited extent. The end-over-end motion was effected by a crank handle and bearings on one side of the barrel, the whole of which was contained upon a wooden framework to raise it clear of the floor. This churn was not quite so dependent upon diaphragms to produce the butter since the tumbling action was itself sufficient. However, diaphragms were provided and could be removed for washing through one end of the barrel which came away completely with the release of some iron clasps. It was normal to provide between 40 and 45 r.p.m. at a temperature of 60 degrees Fahrenheit to make butter, but this could never be guaranteed and on difficult days a single churning may have lasted for several hours. When it was finally made, and that could be ascertained by sound or visual confirmation through a glass-covered peep-hole on one side of the barrel, the butter was doused in cold water to harden it, and the churn then gave a few more revolutions. The buttermilk and the water was next drained away through a tap in the base of the barrel, and the butter left inside the churn was washed several times with clean water before being given a final wash with a solution of brine to aid preservation. The butter was removed from the churn and placed on the **butterworker**, where any remaining water was totally removed from it. The barrel churn, the lid and the diaphragm were finally scoured with boiling hot water and blocks of salt before they were put away.

Box churns were commonly known as tub or washer Churns, the smaller ones being hand-operated and the larger ones driven by mechanical means. The shape varied a good deal despite the name 'box'; some were square, some rectangular, and others cylindrical, but no matter what the shape the distinctive feature of the box churn was the radial paddles that revolved inside the container.

The size of this churn varied to suit the requirements of the dairy but a wooden container 18 in. long, 18 in. deep and 12 in. wide or thereabouts

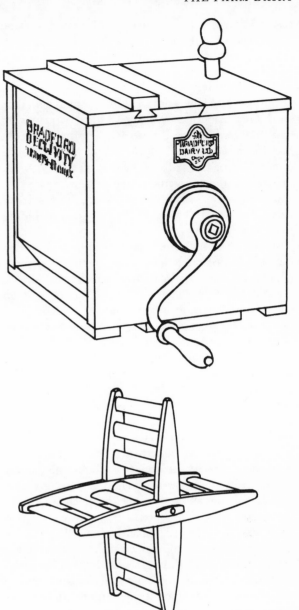

Box churn

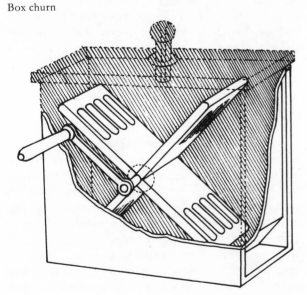

In the box churn the dashers revolve with the crank handle. The essential difference in the various box churns was the shape of the dashers

would have been adequate for the average farm. The radial paddles that agitated the cream were also made in wood and radiated from a central spindle turned by a crank handle. Each paddle was perforated with small holes or formed with thin slats so as to interrupt, but not obstruct, the free passage of cream as it was being churned. This was achieved by rapidly turning the crank handle and consequently the paddles, which caused the cream

227

inside the box to be displaced. Movement of the liquid also caused a current of air to flow through holes in the top of the container, and being forced to pass through the cream, it greatly assisted the making of butter.

Burgess and Key were the London merchants for Anthony's American Washer Churns, which were first imported during the 1850s when they gained a high reputation for quick butter-making. The dimensions of this 14 lb. box churn were: width $12\frac{1}{2}$ in., length 19 in., extreme depth in the centre $12\frac{1}{2}$ in. and length of crank handle 8 in. This size of American Washer Churn cost £2.15s.0d., including three sets of radial paddles.

The cradle or swing churn was simply a wooden box capable of containing 2 or 3 gallons of milk, which was entered through a narrow aperture in the topside. When in use, the churn was lowered into a wooden framework fitted at the base with two wooden rockers, similar to those on a cradle, and, by means of a pole handle, was rocked back and forth at the rate of a clock pendulum until butter was made. The churn was never completely filled so that the milk was free to flow about between vertical wooden slats, and whilst it served the purpose of making butter extremely well, was never fully accepted outside Wales.

The double-action churn was of the plunger or dash-churn variety. It appeared about the middle of the nineteenth century and was immediately popular owing to the manner in which it quickly and efficiently made butter. It consisted of two steam-driven plungers, placed vertically side by side in the same churn. The cream, contained in the churn, was forced through a perforated partition between the plungers by the downward stroke of one plunger, whereupon it met with the downward stroke of the second plunger, and in this way the cream was forced to pass backwards and forwards in rapid succession, which would produce particles of butter in a few minutes. Then the plungers were moved slowly so that the particles could coagulate before the churn was emptied.

Dalphin's American churn first appeared in British farming catalogues about 1845. It was only one of the many effective churns devised by Americans for the foreign market, and this one in particular met with a very favourable reception. In this model, the cream was agitated by patent 'rotary dashers' comprised of wooden slats set at an angle so as to force the cream towards the centre of

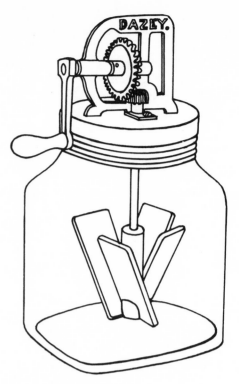

Glass jar type of butter churn

the churn. There it met with 'movable floats' which opened when revolving and directed the cream outwards. These self-adjusting floats and the contrary motion forced upon the cream produced butter in a most effective manner, and expelled the buttermilk from it. Other makers rightly acclaimed its effectiveness and proceeded to build churns on the same working principle. By 1894, there were such a variety of churns and churning powers available that the Royal Agricultural Society held a trial in order to determine their merits. The barrel churn remained much used in farm dairies, whilst churns were gradually replaced by centrifugal separating and butter-making machinery in the butter-making factories.

The illustration shows a churn, the container part of which is a glass jar with a screw lid. They were made in sizes suitable for 1 to $1\frac{1}{2}$ gallons of cream, but were normally filled two-thirds full. They were widely popular during the late nineteenth century but required much churning, by means of the cranking mechanism on the lid, to produce butter.

The mechanical butter worker. Any whey present had to be removed before butter could be weighed and formed into pats. This was formerly done by hand kneading, but no matter how

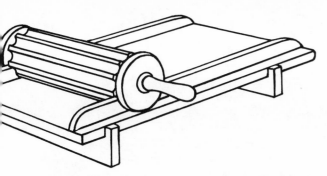

The simplest butter worker rested upon a table top or other suitable support

Butter stamps, roller markers and scotch hands

thoroughly the task was executed some particles of whey remained to cause fermentation and early souring. Thus the price of the butter was considerably lowered. The butter worker had become a very important item in the dairy by about 1870. It was a simple construction yet worked perfectly in expelling every particle of whey, without the produce having to be touched once by hand. It consisted basically of two items, a shallow trough of wood or iron about 24 in. long by 18 in. wide, and a wooden roller 18 in. wide and 6 in. in diameter, which worked backwards and forwards under pressure inside the trough, being turned by a crank handle and guided by tracks along each side of the trough. The surface of the roller was indented with a large number of deep longitudinal or diagonal grooves, which crushed the butter placed inside the trough, and any water that remained after churning was forced out of it and along in front of the moving roller to the end of the trough, where it passed out through a narrow slit. The roller at this point met with stops attached to the track, whereupon the handle was cranked in the opposite direction and the roller proceeded to pass back over the butter for a second time. This action was repeated until the butter was completely free of water and suitable for manipulation by the scotch hands.

Butter hands. When butter was to be formed into rolls or balls, two wooden spades, commonly known as butter or scotch hands, were employed. These hands were identical. The blades were about 6 in. long and 4 in. wide with a handle attached to the top edge, and one side was deeply incised with parallel grooves. The user held one in each hand and used the grooves to roll the butter into shape,

finishing off its surface with a zig-zag pattern, made by the serrated edge of the hand. Butter hands and **butter moulds** are only used now where the dairy work is local.

Butter moulds. Whilst butter was prepared by hand, pictorial images were impressed upon its surface to give a decorative effect. The patterns and images derived from heraldry were most popular, and there might also be an indication of the district where the butter was made. The moulds were in the form of wooden stamps about 2 in. square or 2 in. in diameter, with a handle on one side and the pattern engraved on the other. These moulds, along with a small wooden roller with a pattern repeated around its circumference, were used in most farm dairies before the twentieth century.

A butter press was introduced in 1880 by the Aylesbury Dairy Company, whereupon it was found, by most farmers, to be more practical and expeditious than the **butter hands** they were accustomed to using. An inexperienced person could use this machine and instantly form butter into neat half- or one-pound blocks with a decorative motif impressed on the topside. The small dairy model was built wholly of wooden pieces which slotted together and was a suitable size to stand on a bench or table top. It worked on the principle of a brick or tile press, with the butter being compressed into a rectangular chamber in similar manner to clay. The complete block of butter was stamped and ejected out of the chamber by the action of a lever which caused the floor to rise up. The motifs stamped into the butter were supplied along with the press and were designed to the purchaser's requirements. Butter is still pressed into blocks but unfortunately without any decorative effect upon its surface.

Cheese-Making Utensils and Appliances

A wide variety of cheeses were prepared in English farm dairies before the twentieth century, by which time factories had generally taken over and relieved the farmer's wife of this profitable chore. The wide variety of flavours were due to the milk coming from different kinds of cows, but more so to the manner in which the cows were fed and the cheese was prepared. The making process was similar throughout the country, but varied somewhat in recipe. G. H. Andrews, in his book of *Modern Husbandry*, 1853, described cheese as 'consisting of the caseous matter of milk combined with a great portion of the oily or creamy part; that is if it be good and rich, which it will not be if the milk be subjected to any skimming process to rob it of its cream, as is the case in Suffolk, and other localities, where it is said a cheese, by having a hole made in the centre becomes a very good grindstone'. This authority chose to illustrate the making of Cheshire cheese, as he said it was 'highly esteemed, not only in England, but all over the world, wherever it can be carried in a sound state'. He describes its making as follows:

The milk is placed in a large tub or metal kettle, and warmed either by putting the whole on the fire, or heating a portion of it, so as to bring the whole up to about 90°. The rennet is then added, being placed in a small muslin bag, and the milk evenly stirred about with it; in about two hours the curds will be formed throughout. The dairywoman then commences getting the curds together with her hands and separating it from the whey until the whole mass is in one lump; this is then put into a cloth, placed in a wicker basket, and is put under the machine and pressed, or it is worked with the hands and knees, the lump being frequently cut with a knife into pieces, and then laid together again, and repressed and cut again, until the whole of the whey is squeezed entirely out. Sometimes a deep vat is used to press the curd in, instead of the cloth; but whatever plan be adopted, the end to be obtained is the same, that is, the thorough extraction of the whey from the curd; for if this be not done, the cheese will cut wet, be full of holes, heave, and be altogether of inferior quality.

The curd being pressed into a hard and comparatively dry lump, is now placed in a pan, and the process of breaking down has to be executed; this consists of breaking up the curd into small pieces, and working it with the hands until it is of one quite even consistency; this is the heaviest work of the process, and must not be carelessly done, or inferior cheese will be the result: machines called curd-breakers have been introduced, and are generally used in some localities; but with the Cheshire dairymaids they are not in favour; they consider that the breaking by hand is in every way superior.

The curd being broken down, and the salt added to the satisfaction of the operator, is now placed in a wooden mould called a cheese-vat. The mould is made of the size of the intended cheese, the curd is placed in it, and a tin hoop, called a girth, placed round the top; just within it, the curd is piled up above the height of the cheese-vat, so as to fill the tin hoop, and even be piled up conically, above that, or a cloth is sometimes used to pack the upper part of the cheese-vat.

The mould is now placed in the cheese-press and subjected to the action of its weight. Cheese-presses are made in a variety of ways; the most common are arrangements of levers which may be so regulated as to cause ever so much or ever so little pressure on the cheeses, they are now made of iron and of convenient form.

The vat full of curds being pressed runs off all the remainder of the whey that had not previously been extracted; this liquid is called thrutchings, and is very different to the first whey: it is eaten by members of the farmer's establishment, and is considered a wholesome and agreeable food (that is, by those who are used to it).

The cheeses remain a day or two in the presses, and the curd in the tin hoop is with the hoop pressed down into the cheese-vat, and a cheese thus formed the size of the vat or mould. It is now carried to the cheese-room where it is placed carefully in the racks, or on the floor: many persons think that cheese is better laid out on the floor than it is if put on shelves in a rack. As the Cheshire cheeses are generally large, and when they leave the presses they have not become sufficiently solid to keep their shape, it is necessary to wrap linen bands around them; this prevents their bulging out at the sides and cracking, which if it took place would very much injure them both in quality and appearance. The cheeses now require constant turning and examining, once every day at least, until the bandages are removed: they are then, with a dull-edged knife, rid of any roughness that may appear on their surfaces, and little damages or cracks stopped up, and afterwards they are well rubbed occasionally with coarse cloths to give them a polish; a grey-blue mould, or bloom, will appear upon them afterwards, all over alike, and give that particular look which is impossible to describe, but which is well understood by those who have had experience in the manufacture of them.

Cheshire is celebrated for its excellent cheeses due not only to the nature of the soil, but also to the careful manner in which the process of cheese making is carried on. The farmer's wife generally makes all of the cheese herself; and, as the profit or loss upon the farm will chiefly depend upon the manner in which this is performed, there can be no doubt but that the greatest possible care will be exercised in doing it; for although the manufacture of Cheshire cheese, as I have described it, appears the simplest operation possible, yet, in its practical execution, there are many points of detail that require great experience to properly carry them out; and in the same locality, many women make much better cheeses than their neighbours, all circumstances being the same.

Cheshire cheese is much finer in quality when the cheeses are made large, it is therefore necessary that a cheese dairy-farm should be of sufficient size to carry at least twenty cows giving milk.

It is of great importance to get cheese dry as quickly as possible, and sent to market, as it loses considerably in weight.

Rennet. The substance with which a dairy farmer converted the milk into curds and whey was usually a calf's stomach tied up in a muslin bag. A calf slaughtered before it was three days old was considered superior, and those obtained from Ireland were especially prized. Rennet has been prepared from chemical and vegetable sources.

Oblong cheese vats replaced a circular variety about 1880. First introduced by Cluetts of Tarporley, they were about 7 ft. long, 3 ft. wide,

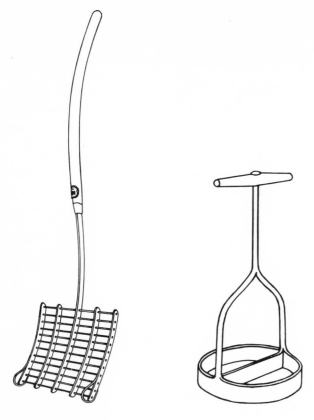

Curd agitator
Curd cutter

This cutter was formed by a circle of copper $\frac{1}{4}$ in. thick, 9 in. in diameter and 2 in. high, which was braced across its diameter by one or two strips of copper of the same height and thickness. This arrangement was joined, at two opposite points of its perimeter, by a forked copper rod which stood 18 in. high and was furnished at the top with a wooden handle. There were many patterns of curd cutter, the best of them made in steel and well tinned to withstand the effect of acid in the **rennet**.
The curd mill could be used in addition to the cutter if the pieces were not sufficiently broken down. This machine appeared in about the year 1830 and remained useful for many decades in the farm dairy. A wooden hopper about 18 in. square and 12 in. deep was suspended over a tub by means of four bearers fixed to its base. The hopper tapered in towards the bottom to leave an aperture 8 in. square, across which was positioned a roller, studded with wooden or iron pegs $\frac{1}{2}$ in. square. These were driven into the roller in rows, $\frac{1}{2}$ in. apart and each peg projected for $\frac{1}{2}$ in. This roller was cranked by a handle and the revolving pegs passed closely between other rows of pegs attached to bars in the bottom of the hopper. When the hopper was filled with broken curd, the operator turned the handle and also pressed the curd down with his free hand, thus forcing it to pass between the revolving pegs, where it was crushed. Cleanliness was essential with all aspects of dairy work, so this machine was made to slot together and could easily be dismantled for washing.

Towards the end of the nineteenth century the curd mill was made in galvanised iron and had become much more compact. The hopper and the essential moving parts were contained as a single unit which could be mounted on a four-legged stand. The curd mill is still a very important item in all cheese-making processes and its operative parts are still much the same as those described.
The chessart, or cheese vat, was made in wood, and its function was to contain curds under the pressure of a press whilst allowing the whey to drain away. Elm wood was commonly used for the construction, being cut into stout staves, sometimes 2 in. thick and bound with iron hoops. The diameter and height of the chessart decided the diameter of the finished cheese, and its capacity varied with the type of cheese being made. It was always open at the top, and the base was also open. If a bottom was provided, then it was pierced with holes through

19 in. deep and contained 172 gallons of milk when full. The inner pan was simply a lining so that hot water or steam could be emitted between it and the outer body for the purpose of heating milk for cheese-making. Whey was drained off from the curd through a tap on one end of the vat, and the curd manipulated by hand.

The curd agitator was often made up to 5 ft. long, with a wooden pole handle and a shovel head formed with brass or iron bars. The illustrated tool had a slightly bowed handle in beechwood. The cutting head, about 12 in. square, was formed with sharp-edged strips of brass, and brass wires, crossing each other at right angles. Similar agitators are still employed.

The curd cutter. When the curd was sufficiently coagulated to be cut, a long sharp knife was drawn through it in various directions. This was followed by the use of a cutting instrument known as a curd cutter, which further reduced the size of the pieces. It enabled the user to reach the lowest strata of the curd and by twisting, lifting, and plunging motions, complete a full fragmentation.

which the whey could drain. Chessarts were used for as long as dairies employed the cheese press, and later ones were often in cast-iron.

The cheese press. It is necessary to apply a regulated pressure in order to remove unwanted moisture and to consolidate cheese. Early devices invariably employed the weight of large stones piled upon one flat stone which rested upon the top of the cheese. Stones were still used to supply pressure during the eighteenth century, but by that

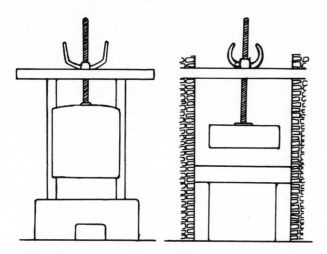

Eighteenth-century stone cheese presses

time they were fashioned into regular blocks and arranged within a simple wooden framework which held the pile secure. No mechanical means had appeared for raising or lowering this mass; it was simply manhandled into position piece by piece. Towards the end of the eighteenth century two separate principles for applying pressure were introduced. The first one, based upon the movement of the printing press of that time, involved the use of a long vertical iron screw, the lowest end of which passed through a large block of stone, whilst the upper end passed through a rigid beam, where it was fitted with a turning device for raising and lowering the stone. The other principle was that of a vertical rachet which was moved up and down by a crank handle and toothed wheel, causing the stone to rise and fall with it. These two mechanical means were in evidence before the turn of the century, whereafter they were exploited by manufacturers along with a system of weights and levers to apply a measured amount of pressure to the curd.

The curds for cheese were put into a stout open-ended container called the **chessart**, and both ends were covered with muslin before it was placed in its position on the base stone of the press directly beneath the weight. Before the weight was applied, a slab of polished stone or wood, called the sinker, was placed on top of the cheese. Its top surface projected up for a few inches above the edge of the chessart so that when pressure was applied, the entire weight of the stones fell upon this slab, which in turn pressed down upon the curd. Any whey that was present was forced out through the bottom of the chessart and escaped through tiny channels incised into the base stone of the press. The cheese press had been replaced to a large degree by the cheese-making apparatus before the end of the century. The latter occupied less space and completed nearly all the work required for making cheeses.

The double-chambered cheese press came into use during the latter half of the nineteenth century and worked in a manner identical to that of the lever press. Some manufacturers employed the screw device, which for some years had been out of favour, but it was re-employed and fitted with a system of weights and chains to apply a continually increasing pressure in the manner of the lever-press. It was often built with two iron **chessarts** about 2 ft. in diameter placed one above the other, which enabled the press to cope with twice the amount of curd as before.

The lever cheese press, made during the late nineteenth century, was a flamboyant cast-iron structure standing higher than a man. The pressure block was normally raised by a ratchet and toothed wheels in preference to a screw. The feature of this press was the long lever arm furnished with a weight which caused the pressure block to descend gradually as the cheese consolidated. The lever was gradually pulled back down to a horizontal position by the presence of the weight and there the mechanism was arrested against movement, unless it was raised once again and further pressure applied to the cheese.

The cheese-making apparatus patented by a Mr Keevil about 1850, subsequently proved to be an invaluable item in the dairy. It effectively speeded up the preparation of curds with its simple mechanism embracing three separate aspects formerly done by hand: breaking up, separation and pressing. It took a circular form, as shown in the

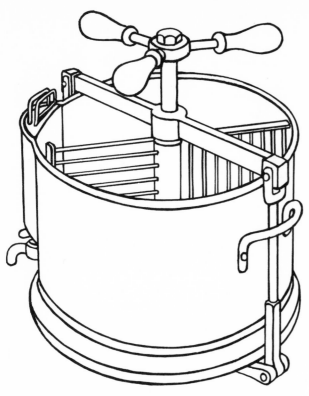

Mr Keevil's cheese-making apparatus

illustration, and was made of brass or zinc of a suitable thickness to withstand pressure. The curd cutters below the centre beam were rotated after the curd inside the container had set, and quickly reduced it to small pieces, whereupon further rotation of the cutters would stir the curd until the necessary degree of firmness had been obtained. The radial cutters and their spindle were then removed and the curd allowed to settle. At this stage the whey was drawn off through taps in the side of the container. Next a brass plate with a number of perforations in its surface was laid upon the surface of the curd and a long screw turned down through the bearing in the centre of the beam, where the cutter spindle had previously been. The bottom of the screw was located into, but did not pass through, the centre of the brass plate, and constant screwing by means of a handle at the top brought pressure to bear upon the curd. This pressure was sustained until the cheese was ready for removal. This apparatus could do nearly all the work required for making cheese, consequently its use became widespread. All later models were based upon the one described, some of them able to cope with 150 gallons or more. They have

since been employed in dairy factories, whilst the small cheese producers retained some version of the cheese press.

Mr Blurton's tumbling cheese rack relieved the dairymaid of one chore, that of daily turning over a large number of newly made cheeses. This inventor lived near Uttoxeter and introduced his idea about 1860, whereupon it was quickly installed in most places where cheese was made in quantity. Until that time cheeses taken from the press were wrapped in cloth and placed along shelves in the cheese loft, where a cool temperature was maintained. In this situation the cheeses would require to be turned over once daily during the first two weeks, every other day during the third week and then once every week until the cheeses had fully matured. This task of hand turning perhaps many hundreds of cheeses daily, was relieved by Mr Blurton's invention, which could accommodate a large number at one time and would reverse them all instantaneously. His tumbling rack was supported within two vertical timbers, each 6 ft. 6 in. high, which were held upright by cross feet or attachment to both floor and ceiling. The shelves to contain the cheeses were formed by twelve horizontal timbers, 14 in. wide by 7 ft. long, and each shelf was placed 6 in. apart in order to allow adequate ventilation. They were braced at the rear edges by vertical slats of wood, which were spaced at intervals and attached to each shelf by a nail. The shelves were tenoned at each end into vertical wooden members, which were provided with an iron gudgeon pin at mid-height. These gudgeons passed through the outer frame members at mid-height, which caused the bottom to be raised up a few inches from the floor, and the whole rack was then capable of revolving end-over-end. All that was necessary to upturn the cheeses when they had been placed upon the shelves was to tilt the rack backwards through 180 degrees, using the gudgeons as pivots so that the shelves faced in the opposite way. By this action the cheeses would settle down upon the shelf which was previously above them, and the rack was then locked in a vertical position by an iron catch until the next day.

Early Milking Machines

Vacuum milking machine, 1862

Mechanical milking. Attempts were made to develop a milking machine during the 1850s, but none was successful until an American, Colvin, produced his hand-operated vacuum model in 1862. The patent was purchased by a Birmingham firm for £5,000, with royalties going to the inventor, and although trade was reputed brisk and the machine awarded a prize medal, nothing further became of it. Colvin's machine was as illustrated. The four rubber teat cups were connected to the vacuum chamber and interior of the pail, and when the cow's teats were dropped into them and a vacuum created by rapid movement of the two handles, the udder was emptied almost instantaneously. But this and other early vacuum machines subjected the cows to unnecessary cruelty, since blood was frequently drawn with the milk. The Agricultural and Dairy Societies then offered prizes for a more practical invention which resulted in a flood of ideas put forward from many countries; most of them dispensed with the vacuum and instead imitated the action of hand milking. They were called lactators.

The Crees lactator of 1881 was one such machine which obtained milk by the action of tiny rollers mounted on revolving belts. It was slung beneath the cow and gripped each teat between a pair of adjustable rollers. As the belts were caused to rotate by a hand crank, so two rollers came together and applied gentle pressure down the length of the teat, removing milk into a container below. There were other lactator machines, but the majority were considered unhygenic and failed to become popular.

William Murchland of Kilmarnock furthered the cause of vacuum milking when he patented in 1889 the first of many machines to come out of Scotland. It was designed for permanent installation, and used a column of water to create the vacuum and a system of 1 in. iron pipework around the inside of the shed, with a branch pipe leading off into each cow's stall. They terminated in rubber-lined teat cups and all milk drained off by the vacuum was collected into **churns**. Murchland's machine worked well, but continuous suction was still recognised as harmful to the cows and was not accepted by many as the real answer.

Dr Alexander Shields of Glasgow presented his pulsating 'Thistle' milking machine at the Darling-ton Royal in 1895, where it gained a silver medal. Whereas all former vacuum attempts had maintained the suction at constant pressure, Dr Shields included a device in his machine which caused a brief but regular punctuation. In this way he provided the teats with some relief, similar to that of the natural sucking action of the calf. His pulsator was to become an important feature of later developments but criticism was levelled against the Thistle as it required too large a pump, the vacuum was difficult and expensive to maintain, and the pipework was complicated. Its cost was high in comparison to other machines, which put it beyond the reach of most farmers. But the principle of this machine was retained and perfected in the twentieth century, notably by Lawrence and Kennedy of Glasgow, and J. & R. Wallace of Castle Douglas. The modern milking machine is much the same, consisting of a vacuum pump, pipeline and a series of teat cups with pulsators to ensure the intermittent suction. They are now to be found installed on all farms where cows are kept in number and only in the remotest areas is milk obtained by hand.

Bibliography

ANDREWS, G. H., *Modern Husbandry* (1853)

The Annals of Agriculture (1783, et seq.)

ARDREY, R. L., *American Agricultural Implements* (Chicago, 1894)

BAXTER, John, *Agricultural Annual* (1836)

Library of Agricultural Knowledge (1837)

BLITH, Walter, *English Improver Improved* (1653)

BOND, J. R., *Farm Implements and Machinery* (1923)

BURKE, J. F., *British Husbandry*, 3 vols., 1834–40

CLARK, R. H., *Development of the English Traction Engine* (1960)

DICKSON, R. W., *Farmer's Companion* (1813)

Practical Agriculture, 2 vols. (1805)

ELKINGTON, Joseph, *An account of the . . . mode of draining land . . . practised by J. E.* (1797)

ERNLE, Lord, *English Farming Past and Present* (4th ed., 1927)

The Farmer's Magazine (1800 et seq.)

FREAM, W., *Elements of Agriculture* (1892)

FUSSELL, G. E., *Farming Technique from Prehistoric to Modern Times* (1966)

The Farmer's Tools: 1500–1900 (1952)

HARTLEY, M. and INGLEBY, J., *Life and Tradition in the Yorkshire Dales* (1968)

LONG, James, *British Dairy Farming* (1885)

LOUDON, John C., *Encyclopedia of Agriculture* (2nd ed., 1836)

MAXEY, Edward, *A New Instruction of Ploughing and Setting of Corn* (1601)

Mode of Drainage (Board of Agriculture pamphlet, 1801)

MORTIMER, John, *The Whole Art of Husbandry*, 2 vols. (1716)

MORTON, J. C., *Cyclopedia of Agriculture* (1856)

PASSMORE, J. B., *The English Plough* (1930)

The R.A.S.E. Journal (Royal Agricultural Society)

RUDGE, Thomas, *The Agriculture of Gloucestershire* (1807)

SCOTT, John, *The Text Book of Farm Engineering* (1885)

SINCLAIR, Sir John, *The Code of Agriculture* (1817)

Systems of Husbandry (1814)

SLIGHT, James, and BURN, R. S., *The Book of Farm Implements* (1858)

SOMERVILLE, Lord John, *On Sheep and Wool, etc.* (1809)

SPENCER, A. J., and PASSMORE, J. B., *Handbook of the Collections Illustrating Agricultural Implements and Machinery* (Science Museum publication, 1930)

TULL, Jethro, *Horse Hoeing Husbandry* (1733)

New Horse Hoeing Husbandry (1731)

WHEELHOUSE, Francis, *Digging Stick to Rotary Hoe* (1966)

WHITLOCK, Ralph, *A Short History of Farming in Britain* (1966)

WILSON, John N., *Rural Cyclopedia* (1849)

The Yearbook of Agriculture, 1960: Power to Produce (U.S. Department of Agriculture publication, 1960)

Trade Catalogues

The author is indebted to the following for much information and the loan of illustrated catalogues, etc.:

Bamfords Ltd.

A. C. Bamlett and Co. Ltd.

Blackstone and Co.

J. I. Case Co., U.S.A.

James Goddard

Howard Rotary Hoes Ltd.

International Harvester Co. Ltd.

Ransomes Ltd.

Savage and Co.

John Shearer and Sons Ltd., Australia

Index

237